CLASSIC WRISTWATCHES 2014–2015

The Price Guide for Vintage Watch Collectors

BY

Stefan Muser

Peter Braun

Michael Ph. Horlbeck

ABBEVILLE PRESS PUBLISHERS

New York London

TABLE OF CONTENTS

A. Lange & Söhne ... 30
Angelus ... 34
Audemars Piguet ... 36
Baume & Mercier ... 40
Breitling ... 42
Cyma ... 48
Doxa ... 50
Eberhard & Co. ... 52
Enicar ... 54
Eterna ... 56
Excelsior Park ... 58
Fortis ... 60
Girard-Perregaux ... 62
Glashütte ... 64
Hanhart ... 66
Heuer ... 68
 Heuer Carrera: Modern Times ... 70
IWC ... 74
Jaeger-LeCoultre ... 82
Junghans ... 88
Lemania ... 92
Longines ... 96
 Longines Chronographs: The Unrecognized Master ... 98
Minerva ... 106
Movado ... 108
Omega ... 112
 Omega Speedmaster Professional: The Legend Lives ... 114
Officine Panerai ... 124
Patek Philippe ... 126
 Patek Philippe Nautilus: Beloved Black Sheep ... 128
Piaget ... 140
Pierce ... 142
Record ... 144
Rolex ... 146
 Rolex Daytona: A Day Is More Than Just Twenty-Four Hours ... 148
Stowa ... 160
Tissot ... 162
Tudor ... 166
Ulysse Nardin ... 168
Universal Genève ... 172
Vacheron Constantin ... 176
Vulcain ... 184
Wakmann/Gigandet ... 186
Zenith ... 188
Zodiac ... 191

Letter to Readers ... 4
Collector's Market: Wearable Solutions ... 6
Auctions: Going, Going, Gone ... 14
Technology & History: Automated at the Eleventh Hour ... 18
Glossary ... 192
Masthead ... 200

FABIEN COUSTEAU'S
MISSION31
DEEPER, LONGER, FURTHER
FABIEN COUSTEAU AND DOXA
TWO LEGENDARY NAMES IN DIVING COME TOGETHER AGAIN
DOXA was chosen to provide the official timepieces for Fabien Cousteau's Mission 31 endeavor, conducted on Aquarius, the only undersea marine habitat and lab in the world located in the Florida Keys National Marine Sanctuary.. The 31-day underwater mission sets a new submersion record for human oceanic exploration and DOXA also sets a record for being the longest continuously submerged dive watch in history.
DOXA commemorated the mission by manufacturing a special edition of 331 pieces of the famous orange face DOXA SUB dive watch, 31 of which have spent 31 days under water with Fabien Cousteau's mission team. A few lucky individuals are able to acquire these commemorative watches; available exclusively from
DOXA: www.doxawatches.com
The Mission 31 SUB is a modern titanium cased version of the popular DOXA watch that Jean Michel Cousteau wore. It features a new high-tech, light-weight case and bracelet with the strength of stainless steel. It is a 44mm, handmade automatic Swiss watch developed specifically for divers and features a limited edition commemorative dial celebrating Mission 31.
For more info, go to
doxawatches.com and mission-31.com
About DOXA Watches
In 1967 DOXA created "orange mania" by introducing the world's first orange-faced dive watch equipped with a patented no de-compression dive table bezel. The DOXA SUB was used by the Swiss Elite dive troops, U.S. Navy and issued to U.S. Navy saturation divers in the early seventies. The DOXA Sub was Dr. Clive Cussler's choice for his best-selling Dirk Pitt® action-adventure novels and has been worn by members of the Cousteau family and Ocean Futures team, among other professional divers.
www.DOXAwatches.com
DOXA automatic
MISSION31 PROFESSIONAL
DOXA
1 8 8 9
www.doxawatches.com
(877) 255-5017
Photo of Fabien Cousteau by Carrie Vonderhaar, Photo of Aquarius by DJ Roller
©2013 DOXA WATCHES AND MISSION 31

LETTER TO READERS

The market for classic collector wristwatches is changing and will continue to do so. But on average, even with all the excesses and exceptions upward and downward, the general direction is up, one can say. Naturally, the routine work at auction houses and daily trade on the retail market are focused heavily on the major, big-name brands. But the prices that are now being fetched for pieces speak for themselves.

For the vintage collector, it is most probably worthwhile taking a good look at the somewhat lesser-known watch brands—for instance using the current, updated *Classic Wristwatches* catalog, in which we present about 900 vintage wristwatches, each with a picture, a concise description, and a development forecast.

The estimated prices and forecasts are oriented on the actual market situation in Germany at the time of printing. Some prices making the rounds on the Internet could be different from those listed in this catalog. But those pieces advertised on the web need careful scrutiny. Is the price given the one that was paid, or are we dealing here with what amounts to wishful thinking? The models presented in *Classic Wristwatches* are, with few exceptions, ones that were actually auctioned at the Dr. H. Crott auction house.

For the sake of accurate comparisons, all the estimated prices and development forecasts are based on very well-preserved, original, functioning, and complete watches, even if the actual watch portrayed and described in detail does not actually meet this standard in some individual cases. Watches that are especially well-documented and come with some of the original accessories (case, presentation box, replacement bracelet or straps, sales label, guarantee certificate, instructions for use, original invoice, etc.) can yield considerably higher prices. The opposite is also true. Models that are in need of restoration or have visible damage may well fetch a price below the estimate.

The prices listed in this book are all based on estimated euro prices. The dollar prices have been calculated from a theoretical and fluctuating exchange rate of $1.30 to the euro. For greater precision, please consult the daily exchange rate.

For the sake of clarity, the models have not been arranged strictly chronologically within a brand, but rather according to function and type. By the same token, the vintage of each model has been highlighted to help establish a chronology.

The little symbols found below the description of each watch are also there to serve as additional identification and for overview purposes. The legend for the pictograms is as follows:

- manual winding
- chronograph
- automatic winding
- alarm
- calendar functions
- certified chronometer

Our "estimated value" is based upon models in top or near top quality that are original, complete, and functional.

- ➚ Increase in value may be expected
- ➔ Good investment
- ➘ Probably valued too highly at the moment

THE COLLECTOR'S MARKET FOR CLASSIC WRISTWATCHES

WEARABLE SOLUTIONS

The iconic designs from the 1960s have captured the headlines in collector circles. But the focus of watch fans has changed somewhat: technical values are making way for more practical style considerations. Naturally, the prices are also expected to go up considerably.

Text: Stefan Muser
Pictures: Auktionen Dr. Crott

The fashion of strapping time to one's wrist is barely a century old. And at the beginning, no one really knew whether wristwatches were for women and dandies or rather for tough guys, like pilots, soldiers, and athletes. As it turned out, the wristwatch managed to cross the gender gap without any problem, and soon after World War I, it was available to everyone and for every taste.

There is an active scene that is particularly interested in these early wristwatches. Their comprehensive collections contain a broad range of anything from rebuilt pocket watches all the way to delicate rectilinear timepieces with new customized movements inside. As an auctioneer, I am often saddened to no end when I have to interrupt an auction due to lack of interest in the room even for a very rare and well-kept piece with an enamel dial and a platinum case—that may have been extremely expensive in the 1920s.

Even the early chronographs are making heavy weather of it in the collectors' scene. The chronographs equipped with pocket watch movements are filigree, technical gems that are for the most part made by craftspeople. But the inherent value of these watches, whose retail price often exceeds the annual income of a minor clerk, is in no relation to the prices being paid out today. Readers so inclined can figure out for themselves what is available from their trusted local watch dealer for $4,000 . . .

BUT CAN IT BE WORN?

One question is dividing the vintage watch scene in two: Can I wear the watch with a three-piece suit or a polo shirt without people looking at me strangely? For those with a historical or technical interest in watches, wearability or suitability of a collector's item for everyday use is of no interest, because they will probably not carry it on their wrist in the first place, but rather keep it at home. But in the ten years during which we have been publishing this catalog, a new group of collectors has arisen within the wristwatch scene, one that is increasingly guided by visual and emotional criteria for purchasing. They are less interested in the technical features of a watch and more in prominent owners, memorable appearances in films, or portrayals in historic black-and-white photographs. And the new vintage collectors would like to wear, show, and look at their little treasures. This attitude also explains the incremental rise in the popularity of modern wristwatches from the

Above: An exotic mix of professional instrument, design icon, and cultural treasure with a lot of history makes Panerai frogman watches highly coveted collector's items. These two rare references were auctioned in the spring of 2013 for about $170,000 (top) and $80,000 (bottom), not including fees.

Opposite page: This early H. Moser & Cie wrist chronograph clearly shows its earlier life as a pocket watch.

so-called era of the "economic miracle" in the 1950s and the style-shaping 1960s, which are back in fashion.

INSIGNIA OF THE HEROES OF THE SILVER SCREEN

Sean Connery's role as James Bond, secret service agent 007, in the film *Dr. No* is unforgettable. He wore a Rolex Submariner on his wrist—first in diver shorts, then in the evening, in a tuxedo. Or Paul Newman's legendary glance at the dial of his Rolex Daytona in the Indianapolis racing film *Winning*. Then there is Steve McQueen in the racing epic *Le Mans* in the role of Jo Siffert driving the Gulf Porsche 917 and dressed in authentic racing overalls with a Heuer patch and a square Monaco on his wrist or Neil Armstrong who tied an Omega Speedmaster Professional over his space suit with a long Velcro band when walking on the moon's dusty surface.

Timepieces of this caliber are favored by high-earning watch aficionados in their prime because they saw them with their own eyes in the movies, magazines, and television programs during their youth and have built up a personal relationship to the past and the events of the past. These watches were the insignia of their heroes.

The fact that many watch brands are reissuing their "legendary" models from the 1950s and 1960s—obviously with contemporary styling, i.e., somewhat bigger and basically always equipped with banal technology—only serves to make the true legends from the old days even more desirable. The offer is quite extensive and affordable, as long as one isn't aiming to acquire that one dial that everyone wants—and at any price.

PESKY DETAILS

Sometimes it seems difficult to understand how a small detail on a Rolex dial can have such an impact on the value of the watch. That value, though, is of a different kind than with a complicated Patek Philippe or an enamel dial of a historic Breguet pocket watch. The version with the special dial was neither costly to manufacture, nor expensive

Rolex Oyster Cosmograph Daytona,
Ref. 6263, ca. 1976,
Caliber 727, manually wound

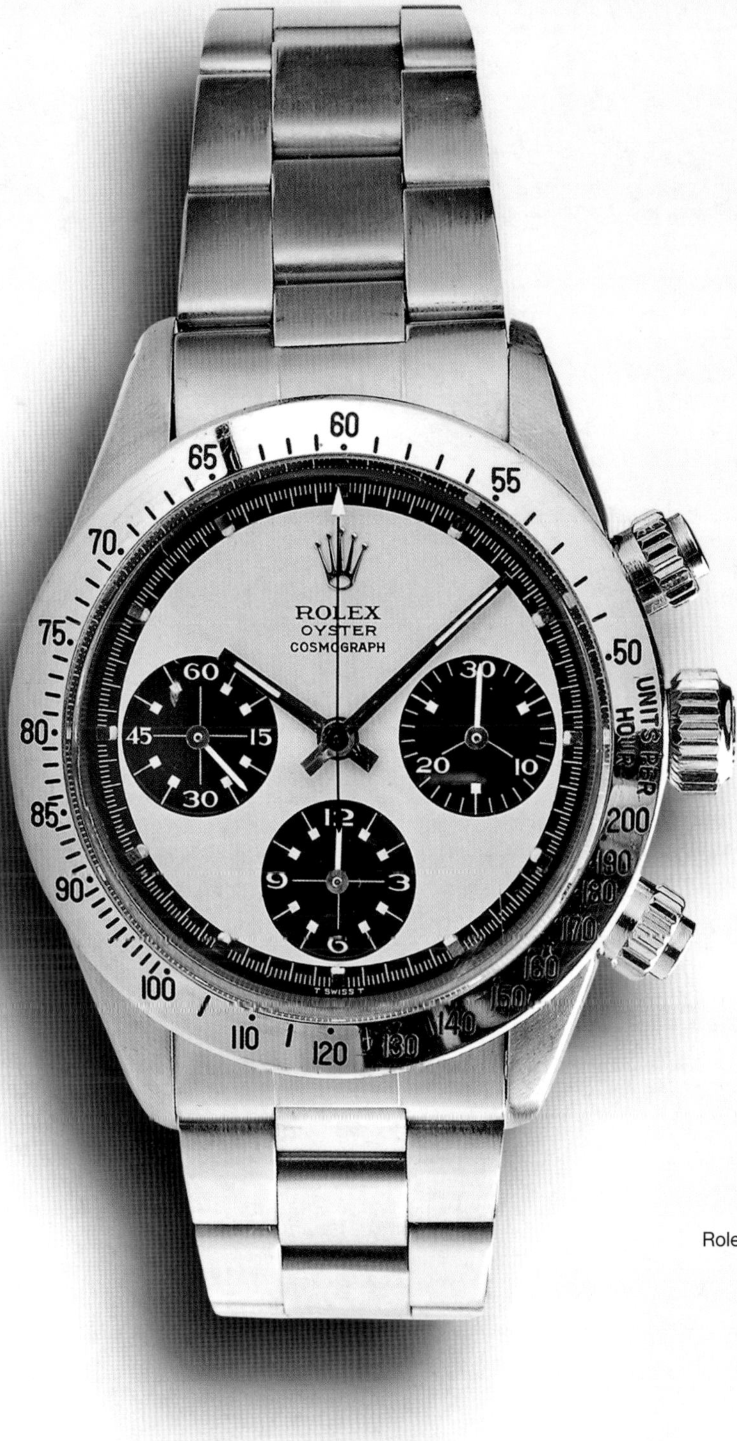

Rolex Oyster Cosmograph Daytona,
Ref. 6265, ca. 1980,
Caliber 727, manually wound

Do you see the difference?
The Daytona opposite was bought in the spring of 2013 for about $19,000. The Daytona on this page went for about $100,000 mainly because of the so-called "Paul Newman" dial with the typical block indexes on the totalizers.

Incomparable elegance, best quality.... but the 34-millimeter case is a little delicate and the watch undervalued. The Patek Philippe with the manually wound Caliber 27-400 is available for up to about $9,000 at this point.

to buy. Only, it was rarer. So, what happens, occasionally, is that a Rolex Daytona with a coveted "Paul Newman" dial, will be overpaid for at $45,000 to $55,000, way above the already robust valuation for an industrially manufactured wrist chronograph with a comparatively simple manually wound movement. I feel that the consumer is no longer buying a watch, but rather a design icon, a rarity, a piece of pop art.

To mock the alleged "spleen" of the Rolex collectors, however, is not very nice. In the final analysis, every watch aficionado has his or her own connection to the material, and access via the visual is just as legitimate as the allure of technology, history, cultural meaning, or value retention. Significantly, it's precisely those "cult" watches techno-fans tend to disparage that maintain their value and at times even evolve into investment objects.

THE REAL THING

Authenticity, historic relevance, and exclusivity are the most important factors for value retention and development. Another good example of this would be the historic Panerai watches, whose prices in recent years have also left the realm of technical reality. Today, auction prices are sought that make no sense if one looks at the watches from the standpoint of classical watchmaking, of the art and craft of the Renaissance or the world of Abraham-Louis Breguet.

The nature of frogman watches is that they were actually war matériel and were involved in military acts. Nevertheless, putting the "Paneristi" in the class of collectors of military memorabilia would be completely off the mark. Panerai watches are witnesses of an epoch, both technically and style-wise, and they convince with their authentic patina, which turns them into unique timepieces. Furthermore, Panerais from the 1930s and 1940s are rare—not as rare as the marketing department of the Richemont Group would like us to believe, nevertheless rare enough to name the individual watches after their first owners. It is the exotic mix of a professional tool and bit of cultural heritage that makes the Panerai Radiomir an icon of design of inestimable value today.

SO WHAT SHALL WE BUY?

Meanwhile there are investment funds that are focused on classic wristwatches. People put money into a common pot and use it to buy watches with the hopes of a solid increase in value in coming years. Of course, they only buy blue chip pieces, a complicated Patek Philippe, for instance, or popular Rolex steel

Another real gem: the IWC Ingenieur sells for up to $5,500.

The price of the Omega Constellation (left) is climbing, thanks to strong demand from China. Popularity of these "modern classics," like this Omega chronograph (right) with a bullhead case, has been on the rise.

watches, and then for a long period do nothing. It seems to be working, because these watches have maintained their attraction over the past years and have distinctly increased in value.

On the other hand, there are losers as well on the vintage watch market. Regretfully, I must say, these include some wonderfully elegant Patek Philippe manually wound watches with interestingly shaped cases and top-quality custom movements. Unfortunately, as far as wearability is concerned, they are far from today's conventional ideals of 1960s style, and that is why their prices never seem to get off the ground. Instead they hover at a level that is far too low, say, around $6,500 to $8,500, for a top-drawer gold watch—a shame, really. But by today's views, they are too small and delicate, and are considered more ladies' watches. For a man to wear a Top Hat or an Hour Glass would require a great deal of self-assurance. On the other hand, this sorry situation harbors an opportunity for genuine connoisseurs. Never have there been so many good watches available for so little money. The same applies to the many well-kept IWC manual winders from the 1950s. Admittedly, compared with a Patek Philippe from the same era, the IWCs appear boring, almost trite. But they are qualitatively well-made—much better than some of the stuff that comes out of Schaffhausen these days. And ultimately, an IWC from the 1950s is distinctly cheaper than a Patek Philippe: in recent times, the prices have hardly passed the $1,300 mark.

Maybe IWC should do more advertising in China. This worked at Omega: a good Constellation cannot be found under $3,000. The fact that this price has doubled in the past five years has to do with the incrementally rising demand in China. Wristwatch devotees there are already gobbling up classic vintage models as the keystone to wristwatch collections. For the Chinese consumer, the primary criteria are a good brand and good looks—function and technical specifications play only a minor role for them. These collectors are naturally focusing on well-known names and intelligible prices. As references, classic Omegas are high up in the hit parade, because Omega already enjoys a great deal of brand awareness with its modern collections. The same applies to the steel Rolexes and early Cartiers.

AND THE REST OF THE STORY?

In my experience, within a fairly short space of time, educated European and American consumers will realize just how undervalued classic 1940s and 1950s chronographs are. Technically and qualitatively, these manually wound movements had already reached the pinnacle of what could be done, and aesthetically, they have not been topped in the past sixty years. This applies above all to the beautiful compax models from Universal Genève and the Longines chronographs driven by the Caliber 13 ZN or 30 CH, watches that are the subject of almost cultlike admiration in collector circles. Admittedly, with their 36 mm cases, these timekeepers are not suited to bragging, and their elegant style does not quite fit in with the sports function of a chronograph by today's metrics. But strictly speaking, these watches are suited to two other areas of deployment, with a suit and with jeans.

The upcoming thing is the classics from the 1970s; that is where the classic watch market converges with the automotive classic market. They attract a younger target group, younger than for the icons of the 1960s mentioned above, but they are still not being hunted that assiduously. Besides, the market spectrum is widening considerably. In addition to the unavoidable Rolex classics, the hot models by Breitling, Heuer, Tudor, Omega, and Longines are on the rise. First and foremost, there are the increasingly robust diving watches being used in extreme depths, as well as the distinct chronographs that were released just before the "quartz crisis." You can read more about this in the following pages, which explore automatic chronographs.

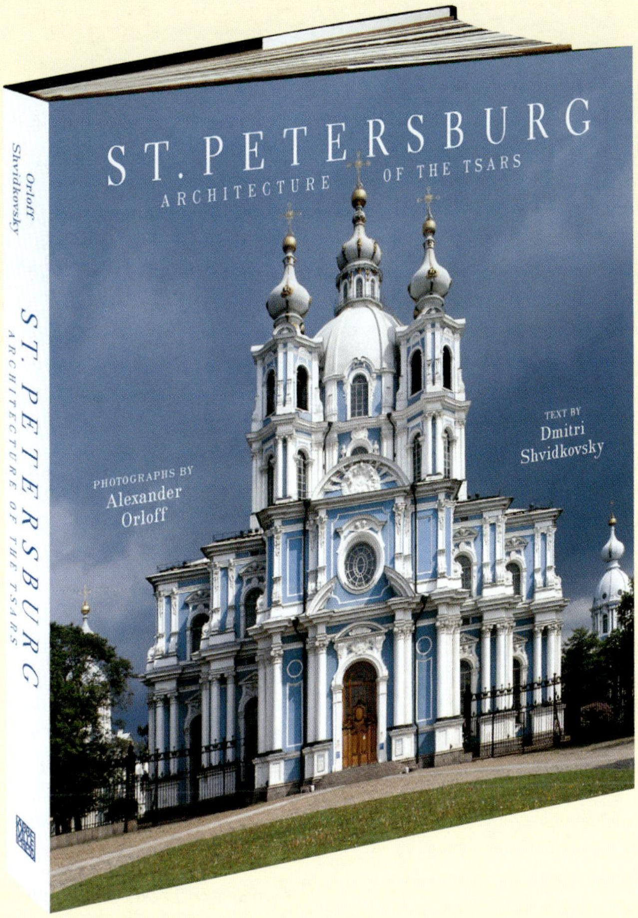

St. Petersburg

Architecture of the Tsars

Before becoming a city, St. Petersburg was a utopian vision in the mind of its founder, Peter the Great. Originally planned as Russia's "window to the West," it evolved into a remarkably harmonious assemblage of baroque, rococo, neoclassical, and art nouveau buildings. Crisscrossed by rivers and canals, Goethe dubbed St. Petersburg the "Venice of the North."

Never before has this city's beauty been captured as eloquently as on the pages of this sumptuous volume. From the stately mansions lining the fabled Nevsky Prospekt to the magnificent palaces of the tsars on the outskirts of the city, photographer Alexander Orloff's portrait of St. Petersburg does full justice to the vision of its founder and namesake.

Text by Dmitri Shvidkovsky
Photography by Alex Orloff
Over 350 full-color illustrations
360 pages · 10⅝ × 12¼ in.
Cloth · $95.00
ISBN 978-0-7892-0217-8

Published by ABBEVILLE PRESS
137 Varick Street, New York, NY 10013
1-800-ARTBOOK (in U.S. only)
Also available wherever fine books are sold
Visit us at www.abbeville.com

THE AUCTION MANUAL

GOING, GOING, GONE . . .

Vintage wristwatches, in particular the valuable ones, are generally bought face-to-face from a jeweler one trusts or a dealer specializing in antique watches. Their inventory is usually limited. An alternative is specialized auctions that boast a wide range of products, flexible prices, personal contact, and a solid business address. Not everyone knows how an auction works, however.

Text: Peter Braun
Pictures: Auctioneers Dr. Crott

Outstanding anchor chronometer pocket watch from the Union Glashütte manufactory in Saxony, Germany, dated around 1900 and featuring the characteristic three-quarter plate movement. Note the gold lever and escape wheel.

The Internet is full of fascinating classic watches, but by the same token, the World Wide Web is far more patient than even paper. There are a number of risks, from the exaggeration of a piece's attributes, to miles of red tape when purchasing a watch from abroad—including customs issues, trying to figure out and pay import taxes, and the legal intricacies that ensue if a sale fails to take place. The old-fashioned method of buying, preceded by a physical examination of the goods and sealed by a handshake or the comforting whack of a gavel, might be safer in the end.

At an auction, it's the auctioneer or the auction house that serves as a guarantor for the sale of the object.

The process is based on mutual trust in two directions: between the seller and the auctioneer, and between the buyer and the auctioneer. This can only work if all parties involved adhere to certain rules.

An unexpectedly hot battle of bidders drove up the price of this Patek Philippe Ref. 1518 to over $1.7 million.

GOING ONCE . . .

The seller will seek information from the auctioneer regarding the sale's prospects and will also state the lowest limit price for a watch. The auctioneer will examine the piece and inform the seller of any damage or problems. This consultation about the watch to be auctioned is also the first test of trust, because any auctioneer concerned about reputation has an obligation to potential buyers to describe the object candidly in the auction catalog. A family heirloom might have been overvalued due to some emotional attachments. The auctioneer will then have to approach the matter diplomatically to avoid hurting the seller's feelings.

Individual experience and expertise are irreplaceable when it comes to uncovering counterfeit pieces or so-called "marriages"—watches that have been pieced together from various models. Ultimately, the auctioneer will have to answer for the transaction with the buyer, even if he or she cannot go after the person who brought in the watch at some later date. It is the auctioneer who is liable for all the information given when the piece arrived, in particular for the descriptions, the physical features, the age, authenticity, and manufacturer. The same applies to the data produced by an expert. The auctioneer is not obliged to check all the information, but it is in the house's best interests to do so in order to avoid any potential disputes with the buyer. The limit price and other agreements between the buyer and the auctioneer are put in writing. That contract will contain terms and conditions, information regarding insurance costs, administration, any repairs or restoration expenses, and, of course, the auction house's commission.

A nonbinding estimate for the sale of the watch can be made using photos and descriptions. Most sellers, however, choose to bring the watch to the auction house to have it looked at and evaluated. In the case of larger collections or estates, the experts will visit the seller on request. After delivery, the watches are catalogued, photographed by professionals, and given precise documentation. The process is quite complicated. But it means that the watches can then be presented in an auction catalog in several languages and on the website of the auction house in a more appropriate manner.

The reserve price is then used as a basis for an insurance policy to cover the duration of the contract, which will extend four to eight weeks after the auction as a rule. Watches

The auctioneer's pulpit: to the left, the bank of phone bidders' assistants; on the right, display cases containing the objects heading for the block.

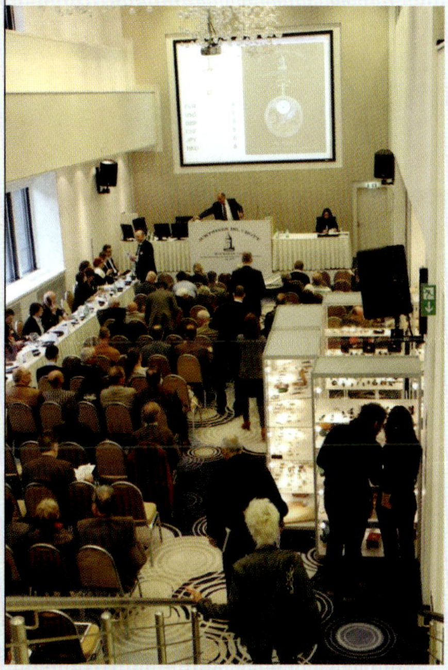

During the auction, the latest bid is projected onto a screen.

that fail to sell must be picked up on time or released for another sale.

GOING TWICE ...

An international watch auction not only displays a wide selection of timepieces from different eras; it is also an attractive backdrop for collectors, dealers, and experts to get together for an exchange of views. Before each auction, the potential buyers will have the opportunity to take a closer look at the objects heading for the block and find out more from the experts and auction house employees present. This will take place either at the auction house's headquarters (usually by appointment) or where the auction itself is to be held. These viewings are usually carried out a day before the auction. In addition to the legally binding description, one can also get information on the object of desire through a personal examination.

Bidders who want to take part in the auction will have to register with the auction house's office and present some form of legal identification. They will then be given a bidder card with a number. Bidding during the auction is done by raising the card. The objects are usually presented in the same sequence as in the catalog and projected on a screen behind the auctioneer—along with the latest bid in various currencies.

A bidder who prefers anonymity or is prevented from coming to the auction for some reason can also send in his or her highest bid. It will be dealt with like any offer in the room; that is, the absentee bidder will

Value in sets: Matching lighthouse clocks like these can fetch large sums especially if they are in in mint condition.

continue bidding until his or her own highest mark. Phoning in a bid is usually only possible for objects of a certain value, say a base price of $1,500, due to the additional costs for the auction house—an employee will have to be in touch exclusively with the bidder throughout the auction. Dr. Crott auction house also offers the possibility of making online bids in real time that are treated the same way as room bids or telephone bids. The bidder has to register for this service in the usual way, with ID.

. . . GONE!

The price you pay for a watch is crucial to the value of that watch—which is why it is wise to read the fine print on the auction contract and set a limit that includes all the fees. The actual purchase price is made up of the "hammer" price plus anywhere from 20 to 25 percent in fees. Sometimes VAT is added as well. When the gavel falls for the third time on the highest bid, it signals the legal completion of the purchasing contract. The bidder can no longer get out of it without legal consequences.

If the seller's minimum price was not attained, the object will be declared sold under reserve. To complete the sale, the seller will have to approve it. Once the bid has been accepted, the object can be paid for in cash or with a certified check and then carried away. In all other cases, the buyer will be invoiced after the auction, either immediately or by mail. Once the invoice has been paid in full, the buyer may either pick up the object at the auction house or have it shipped. Sellers whose objects were successfully sold at the auction will receive an accounting of the process. It includes the final price minus the auctioneers provision and any ancillary expenses. The auction yield will then be transferred or paid by check.

Transparency is key to the whole transaction. The auction house must act as a professional business partner for both the seller and buyer, because in the event of any claims later on, it can be called upon to clarify matters. Transparency means that all participants can rely on a relationship of trust and problems later on are the exception. After paying for the watch, the buyer has the law on his side; a dishonest seller will only be able to try the trick once. In this business, the first time is usually the last as well.

Stefan Muser, auctioneer and owner of the Dr. Crott auction house in Mannheim, Germany.

The auction room at the Sheraton Airport Hotel in Frankfurt, Germany, where watches are put on the block twice a year.

DEVELOPMENT OF THE AUTOMATIC CHRONOGRAPH

AUTOMATED AT THE ELEVENTH HOUR

In the 1960s, the mechanical watch with automatic winding reached a temporary pinnacle. The rotor-driven automatic had proven its mettle in terms of reliability; the winding performance was so good that it could be combined with virtually any complication—except the chronograph.

There were many different mature calibers on the market, but the obstacles to using an automatic movement appeared insurmountable. And then, in the nick of time, the watch industry actually managed to connect an automatic movement with a chronograph—right before the end of the "mechanical era." For watch aficionados, this opened up a small but fascinating collection niche.

Text: Michael Ph. Horlbeck
Pictures: from the archives of ARMBANDUHREN

It was a perfect sensation: When Movado and Zenith jointly presented the first chronograph with an automatic caliber on January 10, 1969, in Geneva, they called it "El Primero," which translates as "The First." It was the final line drawn under one of the most ambitious horological projects of the previous decades. One of the last watchmaker's challenges had been met with success. But Movado and Zenith were not the only brands to have examined the problem.

Nineteen sixty-nine was the year of the automatic chronograph. After Movado/Zenith, it was the turn of Breitling and Heuer. On March 3, 1969, they presented their Caliber 11, developed as a cooperative project. In order to generate enough media attention—having already missed out on being the first out of the gate—the presentation took place simultaneously in Geneva, New York, Tokyo, Hong Kong, and Beirut.

Nevertheless, at the other end of the watch world, things were also moving fast. While the Swiss rivals were all over each other to be the first to develop and present prototypes (because neither of these movements was actually ready for production), Seiko's customers were able, as of June 1969, to acquire automatic chronographs from large scale production. In this regard the Seiko Caliber 6139A was not only ahead of the game chronologically.

INSURMOUNTABLE OBSTACLES

The idea of combining an automatic with a chronograph had already been floated in the late 1940s once the automatic had proven its core functionality. The problems that needed solving at that point seemed far too difficult to even consider implementation seriously. It was no accident that equipping a chronograph movement with an automatic winding mechanism was seen as one of the major construction challenges in watchmaking. A freely turning rotor including all the required winding gears had

Exploded view of the fully integrated "El Primero" chronograph movement with automatic winding via a central rotor

Opposite page:
Left: Breitling Navitimer Chrono-Matic with Caliber 11
Middle: Zenith chronograph "El Primero"
Right: Seiko chronograph with Caliber 6138/6139

ZENITH CALIBER "EL PRIMERO"

Integrated chronograph movement with automatic winding, central rotor

DIAMETER: 30 mm (13¼''')

HEIGHT: 6.6 mm

FREQUENCY: 36,600 vph

JEWELS: 31

SHOCK PROTECTION: Kif

to be mounted over the chronograph cadrature without affecting its function. Since both assemblies needed the same space on the back of the movement, it created a classic dilemma: either the chronograph or the automatic.

The problem could be easily solved, in theory. The two functions had to be arranged on the two sides of the movement. If, as in the case of the classic models, the chronograph cadrature was placed on the dial side, the entire bridge side would be available to house the automatic. This solution, however, conceals two decisive technical disadvantages: To set or adjust the chronograph mechanism, the hands and dial would have to be lifted each time—something unacceptably complicated during assembly, strictly impossible for servicing. The other problem would be the sheer impossibility of displaying the date before or on top of the chronograph cadrature, which is why the chronograph had to be on the bridge side. But even if a winding rotor could be placed above the chronograph cadrature, it would create another problem: height. A normal hand-wound chronograph was built about as high as a simple automatic caliber. The hubless ring rotor had already been discovered, so the actual assembly height could have been reduced with the oscillating weight placed between the movement and the case. But the unsatisfactory winding power made this system unfit as a serious chronograph solution. So some way had to be found that would allow integration of the automatic subassembly into the chronograph caliber. However, engineers were aware that no usable solution could be developed before patents would run out, since the best solutions regarding energy efficiency were legally protected.

Some design engineers were even reconsidering the entire principle of the mechanical watch. The introduction of the Bulova Accutron with its electrical tuning-fork movement raised the fundamental question as to whether the time to develop a mechanical chronograph had passed in the first place. Talk of the day was all about quartz, although this precision technology was also extremely expensive. In 1968 and 1969, no one could imagine that in a mere decade the quartz watch would experience an extraordinary price collapse. And so, in that little

Above: The Zenith Caliber 400, the legitimate successor of the 1969 "El Primero" Caliber 3019, essentially unchanged

Right and Opposite: Two typical Heuer chronographs with the Caliber 11, easily recognizable from the 180-degree switch of crown and pushers. The photo of the movement shows the chronograph cadrature; the microrotor is integrated under the mainplate.

window of opportunity that had opened between doubt and hope, three groups ended up presenting three entirely different solutions based on what was already an outdated mechanical foundation.

THE LONG ROAD TO SIMPLE SOLUTIONS

Even though 1969 was the year of the automatic chronograph, the development had already started decades earlier. The actual pioneer of the automatic chronograph was the movement maker Lemania, which belonged to Omega. In 1946, their technical director Albert Piguet managed to manufacture a functioning prototype of a chronograph with a self-winding mechanism. But an attempt to mass-produce the prototype ultimately failed due to Paul-Emile Brandt's (Omega) pessimism about the product's market chances.

Among the other forgotten pioneers of the automatic chronograph was the Geneva-based brand Universal Genève. It was one of the renowned Swiss chronograph specialists, responsible for the tricompax model, which is to this day synonymous with the typical "3 – 6 – 9" ordering of totalizers on the dial. The in-house chronograph caliber was developed by the subsidiary company Martel Watch in Les Ponts-de-Martel. In the watch industry, Martel was seen as a chronograph expert, but it had also proven its mettle in automatic watches with the development of Calibers 1115 and 1118. Universal Genève felt it would make sense to commission Martel to make a new automatic chronograph. The cost of achieving it appeared immense, but the company Movado from La Chaux-de-Fonds was prepared to collaborate and contribute to financing the development. Movado itself was on the lookout for a substitute for its Caliber M90, an outstanding chronograph caliber based on an existing manually wound caliber. The chronograph module was located on the rear of the movement. Typical for this construction was the otherwise unusual position of the subsidiary seconds dial at 6 o'clock.

But before the first results were obtained, Universal Genève hit a financial reef. The first thing to go overboard was the majority stockholding, next came Martel. The com-

HEUER, BREITLING, AND HAMILTON-BÜREN, CALIBER 11

Modularly built chronograph movement with microrotor-driven automatic winding

DIAMETER: 31 mm (13¾''')

HEIGHT: 7.7 mm

FREQUENCY: 19,800 vph

JEWELS: 17

SHOCK PROTECTION: Incabloc

Strictly speaking, Seiko was the first manufacturer to put an automatic chronograph on the market, but only in Japan, not in Europe.

pany Zenith in Le Locle seized the opportunity and acquired the movement specialist, which had already built the first Zenith rotor-driven automatic, the Caliber 2522. It was a caliber that already contained some of the particular solutions of what later became the El Primero. The collaboration with Movado came to an end in 1966, when Bulova bought out Universal Genève.

Those in charge at Movado decided on a collaboration with Zenith, and so in the end the two companies forged an alliance through the movement specialist Martel, in order to begin jointly developing the complex mechanism. Martel began work on a classic integrated caliber (3019 PHC), with column wheel control, tricompax display (subsidiary seconds at 9 o'clock, 30-minute counter at 3, and a 12-second counter at 6) and an optional perpetual calendar with a moon phase (3019 PHF). One particularly modern feature was doubling the frequency from the then habitual 18,000 vph to 36,000 vph.

Already in 1965, a second group had formed with the objective of developing an automatic chronograph. Nowadays, Jack W. Heuer is considered to have been the driving force. This brand-new majority shareholder and junior boss of the world's largest manufacturer of chronographs recognized the opportunity of doing something to offset declining sales by developing a new caliber. He and Willy Breitling got together to establish a partnership between their two companies. Development costs were to be shared, and the product was to benefit both companies as well. This collaborative effort by two renowned chronograph specialists, so rare for Swiss enterprises, had a disadvantage in comparison to the Movado-Zenith-Martel one in that neither company had any technological know-how in terms of automatic calibers and chronographs. Both Breitling and Heuer were merely assemblers, who picked up their needed movements from Valjoux (Heuer) and Venus (Breitling).

The two men decided to work on a modular caliber that would keep the development within certain controllable boundaries. To keep costs down, an existing automatic, an especially thin one, was chosen as a base. Following the Movado model, it was to be given a module. After some searching around, they settled on the Intra-Matic caliber in an ultra-thin automatic watch made by a company called Büren. It had an additional and decisive advantage, namely an integrated microrotor. So the back of the watch was not entirely filled by the oscillat-

SEIKO CALIBER 6139A

Integrated chronograph movement with automatic winding, eccentric changer, no possibility of hand-winding; column wheel control, vertical chronograph clutch

DIAMETER: 26.9 mm (12''')

HEIGHT: 6.8 mm

FREQUENCY: 21,600 vph

JEWELS: 17

ing mass, leaving at least some space for the installation of the additional module.

The sire of this particular caliber was one Hans Kocher, then technical director at Büren and celebrated as the "King of the Automatic" in the Swiss watch industry. For the development of the chronograph module, the group then looked to specialists Dubois Dépraz in Le Lieu. The company had already made the first chronograph with a cam lever control, the Caliber Landeron 48. The cam lever was more reliable than the column wheel and more efficient to manufacture, so it found its way to the chrono module.

Using the Büren caliber did reduce costs considerably at first, but it also meant compromises when the actual construction began. The structure was supposed to be classical, with the cadrature on the rear of the movement, so it became amply clear that the crown and the pushers could not be arranged on one side. The technicians at Dubois Dépraz then decided to put the crowns opposite the pushers. The reason given later on was that changing the crown's position optically enhanced the automatic watch. The chronograph bridge was shaped in such a way as to ensure isochronous fine regulation at all times without putting components in the way. The other problem that arose with the newly named "Chrono-matic" was the arrangement of the totalizers. Ultimately the rotor axis of the micro oscillating weight had to be drilled through to make space for the spindle for the 30-minute counter. On the other hand, until the introduction of the Caliber 15, customers had to forgo a permanent seconds' hand.

Dubois Dépraz developed a separate module on one of its own mainplates. The advantage with this system was that the module could be delivered to Breitling and Heuer as an assembled piece. It was then installed on the prepared Büren caliber. The manufacturer chosen was Kelek in La Chaux-de-Fonds, which was already back then the "extended workbench" of Dubois Dépraz and well-versed in the manufacturing and assembling of minute repeaters and chronographs. The know-how needed for assembling and adjusting was all there, and it guaranteed smooth production.

RISING SUN

While Movado and Zenith and Breitling and Heuer were basking in the glory of their automatic prototypes, the people at Seiko were already working on optimizing serial manufacturing of the Calibers 6138A and 6139A. Under the radar of the European horological world, Seiko had been working on the design of an automatic chronograph. And the Seiko engineers had gone at it with great efficiency. The cadrature for the chronograph started being developed in 1967 using the 25-jewel Caliber 6160A as a basis. The mechanism was completely concealed under a square bridge. For its Calibers 6138A and 6139A, Seiko adopted a classic control wheel for the chronograph functions, but already when activating the column wheel, one notices the reduction of the functional parts. The positioning of the control wheel above the winding stem simplified operations. Instead of the standard Swiss two-part actuating lever, which usually occupies a third of the movement, Seiko used a short, one-part lever that impacted the control wheel directly. The zero-reset lever and the heart lever were built as simply and, above all, as small as possible.

Seiko avoided the classic horizontal clutch with its very bulky clutch lever and a correspondingly large moving mass. This was replaced by a space-saving and much lighter vertical (axial) friction clutch. The additional advantage of this system was that the clutch lever fixes the position of the second hand as soon as the measuring is interrupted, which avoids the need for an additional blocking lever. The vertical clutch is controlled by two clutch levers that are spring-meshed with one another. One of the two clutch levers is activated by the control wheel and automatically moves the second lever, which creates a scissor-like movement. In terms of functionality, the two clutch levers work like a stop-seconds chronograph. On the other hand, there was no longer any place for a permanent subsidiary second display. Seiko also arranged the dial in a new way.

The minute counter was placed at the 6 o'clock position, the hour counter at 12. For the automatic winding system, Seiko used tried-and-true technology and chose the well-known eccentric changer. Instead of presenting the prototypes to the public with lots of bombast and then working on industrialization and serial production, Seiko began quietly delivering the first Caliber 6139A—but only in Asia. The company did not foresee exporting to Europe or the United States.

PIONEER REDUX

In the latter part of the 1960s, the actual trailblazer in automatic chronographs, Lemania, became active again. Raoul-Henri Erard developed a new chronograph caliber according to Omega specifications. The project was managed by Albert Piguet, who had already

The Lemania automatic was used by various brands; at top is a picture of a rare, high-end BWC chronograph. The movement is the Lemania Caliber 1340 without the winding rotor.

LEMANIA CALIBER 1340

Integrated chronograph movement with automatic winding via a unidirectional central rotor; sweep stop seconds and stop minutes

DIAMETER: 30 mm (13¾''')

HEIGHT: 8 mm

FREQUENCY: 28,800 vph

JEWELS: 22

A rare Citizen chronograph in a typical bullhead case with Caliber 8110A turned 90 degrees

CITIZEN CALIBER 8110A

Integrated chronograph movement with automatic winding; reduction gears; column wheel control, vertical chronograph clutch

DIAMETER: 27 mm (12''')

HEIGHT: 6.7 mm

FREQUENCY: 28,800 vph

JEWELS: 23

completed the first prototypes in 1947. Reliability and functional stability were the top priority, so no effort was made to design an elegant or especially thin movement. A classic chronograph cadrature was used with a modern cam lever control, which took up just about all the available space on the movement side. Lemania solved the space problem in a remarkably simple fashion: For the automatic movement, the chrono bridge was modified in such a way that it could contain the gearwheels needed for the automatic unit. This did mean that the force was conveyed toward the spring barrel at the wrong spot (because the spring barrel was located elsewhere entirely), but the problem could be solved fairly easily by an additional gearwheel. What also facilitated implementation of the solution was the fact that there was no jewel in the chrono bridge and space had to be planned during the design for the minute counter wheel, because the stop minutes were to be displayed from the center.

The chrono bridge had space not only for the automatic's gearwheels, but could also house the rotor bearing. In order to obtain the highest winding performance, Lemania placed a unidirectional rotor in the Caliber 1340, which did not need a changing gear unit.

Another novelty was introduced into the world of chronographs with the Caliber 1040/1340: plastic. No screws at all were used on the dial, for instance. Instead, a small plastic ring was employed as a stable support for the dial and also to affix it. The feet of the dial just had to be stuck into two small holes in the ring. The ring itself was affixed to the mainplate using three small hooks. In the wake of improvements made during the production process, the metal chronograph blocking lever was exchanged for a plastic component. Because the blocking lever is supposed to hug the teeth of the chrono center wheel at the end of each stopping process, it made technical sense to manufacture the lever out of plastic in order to avoid damaging the meshing.

Two versions of the new caliber were manufactured; first, of course, the Caliber 1040 for the parent company Omega, which put it in its new Speedmaster Mark III Automatic. When the caliber was presented to the public in 1970, it was also available for delivery, in contrast to the other two Swiss calibers. The movement bore the name Lemania Caliber 1340 (22 jewels, eccentric fine regulation) and 1341 (17 jewels) and was later used in other brands, in particular the group company Tissot. The most noticeable difference compared to the movements introduced previously was the central minute counter. This allowed for easy readability for each 60-minute span. The Calibers 1040/1340 featured a 12-hour counter and a 24-hour display. In 1973, the Caliber 1041 in the anniversary chronograph Omega 125 became the first automatic chronograph available with chronometer certification.

A WALLFLOWER WITH SPECIAL TECHNOLOGY INSIDE

Almost surreptitiously, the lesser-known company Tenor-Dorly SA in Tramelan and its main movement supplier Breitenbach (Brac) in the canton of Solothurn joined together and began developing a new chronograph caliber. Until the end of the so-called "watch statute," Brac had been forced to produce pin lever movements. Once this limitation dictated by the watch federation was lifted, the company changed over to more modern lever escapements. This meant that it had a solid movement base to construct a new caliber. Even though Brac produced technically mature calibers, the engineers lacked the critical know-how to develop a chronograph. So, they followed the Breitling and Heuer lead and divvied up the tasks. Brac manufactured the base Caliber 1335, a modern automatic caliber with a bidirectionally winding rotor. It was Dubois Dépraz that was commissioned to develop a chronograph module that would fit on the bridge side. Because this caliber featured a large central rotor, the chronograph cadrature had to be constructed in such a way as to find space between the movement and the rotor. To avoid the movement becoming too tall, Dubois Dépraz made a cadrature that was just 2.4 millimeters thick. In spite of the structure's delicate dimension, Dubois Dépraz had decided to control the chronograph using a precise column wheel, but managed to conceal it well under the chronograph bridge. It was worth the effort, and for the next twenty years, Caliber TDB-K remained the smallest available automatic chronograph. In order to ensure rational manufacturing without having to amass their own know-how and expensive controlling machines, another chronograph specialist, Kelek, was brought on board. Kelek was tasked with producing the necessary components, assembling the movements, and fitting them into cases—as with the Caliber 11 of Breitling and Heuer.

The collaboration of the three companies was ultimately reflected in the name of the new caliber, TDB-K. The first three letters

The TDB-K Caliber 1376 has a disc display for the hours (jumping) and minutes. What is unusual is the positioning of the 60-minute counter at 11 o'clock.

stood for Tenor-Dorly and Brac, the K for Kelek. First, it was the turn of Caliber 1369, whose most conspicuous feature was the strange, asymmetrical ordering of the subdials, with the small seconds at 6 o'clock and the minute counter at 10 o'clock. This created an image problem for this caliber. Because of the arrangement of the dial, the Caliber TDB-K could easily be confused with a cheap pin-lever caliber by Ebauches Bettlach. The companion Caliber 1376 had been modified for optional digital display of time and therefore received its own caliber description. Instead of the normal group of hands, two discs with printed hours and minutes were used. To make sure the time could be read accurately, the hours were jumping. But this was not the only change to the movement. The totalizers were also different. The subsidiary seconds at 6 o'clock was sacrificed for the date display, which in turn relinquished its place at 9, since that is where the hour disc turned. The minute counter between 2 and 3 o'clock had to make way for the rotating minute disc. To prevent the Caliber 1376 from becoming a chronograph, a 60-minute counter replaced the hour counter.

The modularly constructed TDB-K caliber has column wheel control; however, the unusually toothed wheel is not immediately identifiable as a column wheel.

TENOR-DORLY TDB-K CALIBER 1369 AND 1376 (DIGITAL DISPLAY)

Modularly built chronograph movement with microrotor-driven automatic winding; column wheel control

DIAMETER: 26 mm (11''')

HEIGHT: 7.6 mm

FREQUENCY: 19,800 vph

JEWELS: 17

SHOCK PROTECTION: Incabloc

Seiko Caliber 7016A

Integrated chronograph movement with automatic winding, eccentric changer, no possibility of hand-winding; column wheel control, vertical chronograph clutch; coaxial 30-minute and 12-hour counters

DIAMETER: 27 mm (12''')

HEIGHT: 6.5 mm

FREQUENCY: 21,600 vph

JEWELS: 17

Even though the finished chronographs were for the most part sold under the brand name Tenor-Dorly, other companies were supplied with TDB-K calibers, including such renowned chronograph specialists as Minerva. A variety of models were even offered bearing the name Kelek.

NEWS ON THE EASTERN FRONT

In Switzerland, four different calibers had already been introduced. Seiko, on the other hand, was the only one present in the Asian market at first. Citizen then introduced the Caliber 8100A there. The Seiko Calibers 6138A/6139A were distinguished by their fine construction; at 5.82 millimeters thick, they were thinner than the Citizen caliber and set the standard for the coming two decades.

Citizen offered its chronograph movement in a version with a 30-minute counter (Caliber 8100A) as well as an additional 12-hour counter (8110A). The engineers at Citizen had also opted for a classic column wheel-controlled chronograph. However, it was entirely covered by the chronograph bridge, or rather, the base for the rotor bearing. So its position could only be identified by the triply slit screw atop the winding arbor. Like Seiko, Citizen also decided to use a vertical clutch, even though it had already had good experiences with its manually wound chronograph. The Citizen 8100A/8110A offered a serial flyback switch. In spite of some special technical solutions, like the seconds wheel installed inside the chronograph bridge, the Citizen movement turned out to be exceptionally thin. Finally, it had a conventional transmission gear (instead of an eccentric changer) and a bidirectionally winding rotor.

THE SECOND COUP: SEIKO CALIBER 7016

Seiko was under pressure, and toward the beginning of the 1970s, it produced a second automatic chronograph movement, Caliber 7016A. But unexpectedly, it turned out not to be an improvement on the Caliber 6138A, but rather a brand-new development that was based on existing automatic movement Caliber 7005A. Although Caliber 7016A had a vertical chronograph clutch, it was fundamentally different from its predecessors—not least due to its smaller dimensions.

This was made possible by profound modifications to the chronograph mechanism, in particular with regard to the column wheel and clutch lever. In contrast to the Calibers 6138A/6139A, not even the control wheel was visible. Rather, the entire cadrature was tucked under the automatic bridge, which means that Caliber 7016A did not even look like a chronograph when seen from the back. With the Caliber 7016A, the double clutch lever used previously was replaced with a one-piece, fork-shaped clutch lever with springing ends. This allowed the springs needed earlier to be left out. As with the classic construction of the split-seconds chronograph, the control wheel of the Caliber 7016A sits between the two clutch levers and controls them simultaneously.

The big innovation with the Caliber 7016A was actually the single auxiliary dial at 6 o'clock. The stop minutes and hours are displayed using two concentrically arranged hands. But reading the time demands some care as the minute counter shows 30 minutes rather than 60. That's why one cannot interpret the stopped time as being the actual time.

THE SWAN SONG: THE VALJOUX 7750

Within a few years, no fewer than a half-dozen automatic chronographs had been made, not counting the derivatives. But only two of these movements reached the open market, and that only with some limitations. The giant movement maker ETA identified the potential of a freely available standard caliber and decided to investigate in-house whether one could combine the tried-and-true Caliber 7733 with an automatic winding system. The designing engineer put in charge was young Edmond Capt, who immediately noticed that the somewhat hoary Caliber 7733 was not very well suited to be extended with an automatic.

So Capt was commissioned with developing a new caliber geared to industrial manufacturing in different versions, which is why he decided to build the caliber systematically, and, as it turned out, very cleverly. By just leaving out the automatic assembly, the caliber could also be used as a manually wound chronograph. In addition, the caliber could come with or without a 12-hour counter as well as an optional full calendar.

To drive the chronograph, Capt chose an oscillating mechanism instead of a clutch lever, since the actual mass that needed moving was smaller. The automatic assembly comprised a special detail. This mechanism, designed for a unidirectionally wound oscillating weight, leaves a lot more space than was needed. One can assume that when designing the basic movement, Capt intended to create enough space for later developments—for example, a

Wine
From Grape to Glass

"The ultimate gift for any wine lover."
—*Touring & Tasting Magazine*

This vividly illustrated text discusses not only awe-inspiring vintages, but also unknown wines from countries only recently included on the wine maps of the world. Half of the book is devoted to the wine making process itself; the other half examines the best wines of the world, country by country, and guides the reader to an understanding of the intricacies of wine tasting and appreciation.

By Jens Priewe
1,000 full-color illustrations
256 pages · 9 × 11⅞ in.
Hardcover · $45.00
ISBN 978-0-7892-0917-7

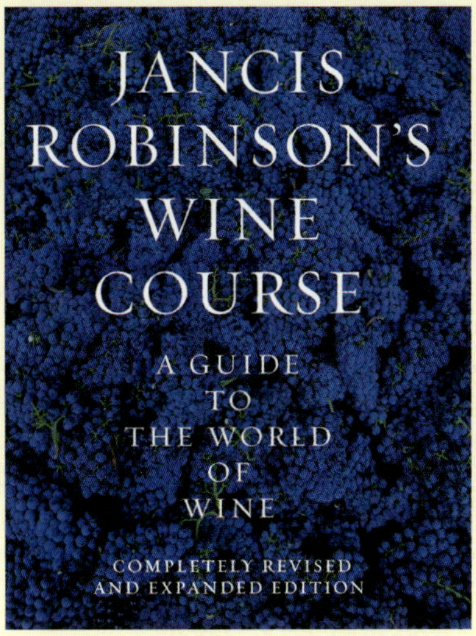

Jancis Robinson's Wine Course
A Guide to the World of Wine

"She has an encyclopedic grasp of her subject and doesn't put a foot wrong . . . a splendid introduction to the world of wine."
—Stephen Brook, *Decanter Magazine*

"Witty, brilliant, authoritative."
—Robert M. Parker, Jr., *The Wine Advocate*

Dedicated to ensuring that you get the most out of every glass, *Jancis Robinson's Wine Course* explains how to taste and store wine, what to serve on special occasions at home, and how to order the best value from a restaurant wine list. Robinson also describes the distinctive characteristics of hundreds of grape varieties and studies the methods employed in the creation of great wines.

By Jancis Robinson
170 full-color illustrations
352 pages · 8 × 11 in.
Paperback · $29.95
ISBN 978-0-7892-0883-5

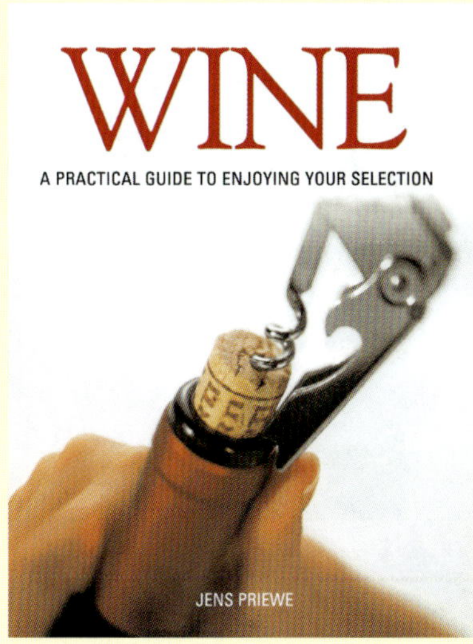

Wine
A Practical Guide to Enjoying Your Selection

Following the success of the author's *Wine: From Grape to Glass,* this book offers a treasury of invaluable information for the contemporary wine consumer who would like to know more about caring for wine and serving it properly. In this essential guide, internationally acclaimed expert Jens Priewe fully describes, with lively text and striking photographs, the correct way to handle wine, from uncorking and tasting to serving and storing.

By Jens Priewe
200 full-color illustrations
128 pages · 7¾ × 10 in.
Hardcover · $32.50
ISBN 978-0-7892-0745-6

ABBEVILLE PRESS

Published by ABBEVILLE PRESS
137 Varick Street, New York, NY 10013
1-800-ARTBOOK (in U.S. only)
Also available wherever fine books are sold
Visit us at www.abbeville.com

Today, the Valjoux Caliber 7750 is produced by ETA (Swatch Group) in large quantities. It forms the basis of many developments by individual brands, such as IWC Schaffhausen. The photo of the movements shows the caliber without the winding rotor.

bidirectionally rotating automatic or a second spring barrel.

What was really new was the use of a computer in developing the Caliber 7750. But Capt did have to go to the mainframe at the Technical Services of Neuchâtel to do the work.

Thanks to the computer, some processes could be simulated, which considerably reduced development times and simplified the construction of a prototype. The Valjoux Caliber 7750 was introduced to the public in 1973, but the "quartz crisis" had already taken hold of the watch industry and so production was stopped in 1975. One little anecdote: it was Edmond Capt himself who was asked to get rid of all the tools needed for production. However, he secretly disobeyed orders, inadvertently ensuring the survival of the Valjoux 7750.

The El Primero's fate was similar. Design engineer Charles Vermot went out of his way to save Zenith's famed chronograph movement, thus safeguarding the survival of the superfast ticker by stashing away plans, tools, and components—and ultimately the survival of the entire company.

Other calibers were not so lucky, and they became victims of the economic tides. In 1966, Büren was taken over by Hamilton and demoted to an assembly operation. As the sales of the Caliber 11 dwindled, the movement was axed. Bulova received the last 10,000 Chronomatic calibers, after which production ceased. The Caliber TDB-K vanished from the market without a sound. Lemania and Seiko were the only brands that continued to produce automatic chronographs, even if the era of the mechanical watch appeared to have reached its end. Back then, no one actually dreamed that a comeback could happen.

Valjoux Caliber 7750

Integrated chronograph movement with automatic winding via a unidirectional central rotor

DIAMETER: 30 mm (13¼''')

HEIGHT: 7.9 mm

FREQUENCY: 18,000 vph

JEWELS: 21 (25)

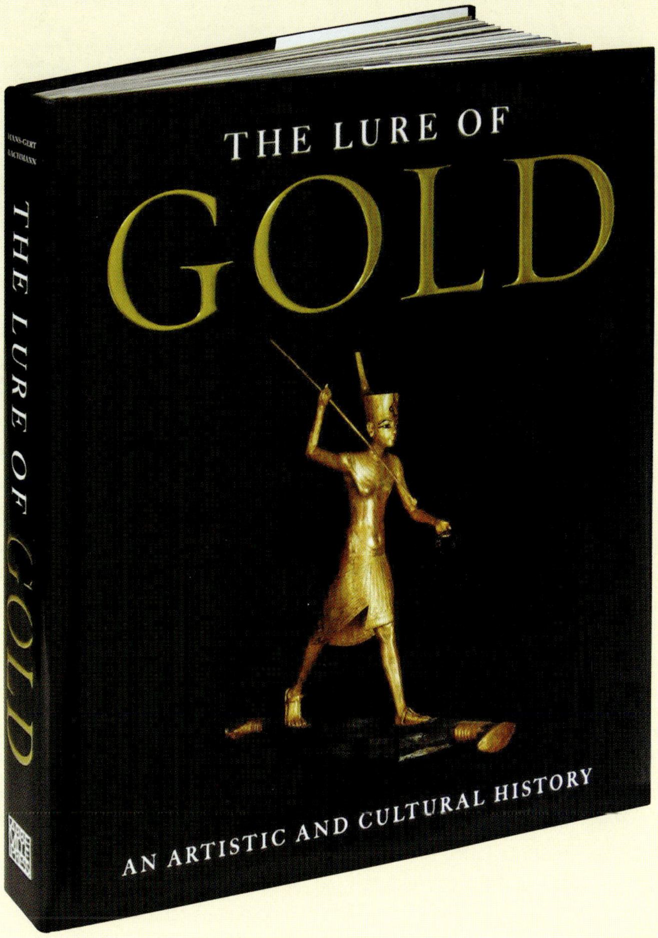

"The book zips through world history as seen in gold's reflected gleam."
—*The Wall Street Journal*

"Pizzaro might have slit your throat for The Lure of Gold. A savvy text relates mankind's venerably irrational mania for the gleaming stuff."
—*San Diego Tribune*

The Lure of Gold

An Artistic and Cultural History

A richly illustrated story of how gold, the world's most beautiful element, has influenced the art, economy, and society of every civilization, this volume explains how people throughout time have mined and refined gold and used it for cultural and economic purposes.

By Hans-Gert Bachmann
285 full-color illustrations
280 pages · 10½ × 11$^{13}/_{16}$ in.
Cloth · $75.00
ISBN 978-0-7892-0900-9

Published by ABBEVILLE PRESS
137 Varick Street, New York, NY 10013
1-800-ARTBOOK (in U.S. only)
Also available wherever fine books are sold
Visit us at www.abbeville.com

A. LANGE & SÖHNE

High watchmaking in Germany is firmly connected to one name: Ferdinand Adolph Lange, who founded the company in 1845 that would spawn the largest hub of fine watchmaking north of Switzerland for several generations.

Lange was born in 1815 in Dresden. After completing his education with court watchmaker Johann Christian Friedrich Gutkaes, he went to Paris for more at the side of Thaddäus Winnerl, a student of the great Abraham-Louis Breguet and the most talented watchmaker of his time. Lange remained in Paris for four years before returning to Dresden, marrying the daughter of his former teacher, and becoming his partner.

The reason for Lange's interest in poor little Glashütte was actually logical: in 1843 Saxony's government called for experts to help alleviate poverty in the Erzgebirge region. They were to found companies there with the help of state start-up financing, thus industrializing the region. Lange received one of the government loans, which he used to hire fifteen young men and train them as watchmakers.

His apprentices had to pledge to work for Lange for another five years after their education ended. Beyond that, however, Lange encouraged them to go into business for themselves and train more craftsmen. Success came quickly with this concept, and Lange was so popular in Glashütte that he was voted mayor and even later a member of Saxony's parliament.

After Lange's passing in 1875, his sons Richard and Emil took over the company. Richard Lange, however, had to retire from daily business in 1886 for health reasons, and from 1919 on grandsons Otto and Rudolf Lange managed the firm.

A. Lange & Söhne manufactured wristwatches in the 1920s, though most of them had Swiss *ébauches* as their base. Before and during World War II, the company mainly produced pilot's and observation watches alongside marine chronometers. Lange pocket watch movements ticked within stately 55 mm cases. The Lange chronometer, manufactured in conjunction with Hamburg Chronometer Works (today Wempe), made the company famous. The marine chronometers' sensationally low rate deviation was only one second every three days.

Production was revived in the dismantled factory after the end of the war. Lange's watchmakers succeeded in setting up the machines for manufacturing the Caliber 28 by 1948. In the same year, however, the company was expropriated, and A. Lange & Söhne was integrated into the newly created combine Glashütter Uhrenbetriebe.

Only with Germany's reunification in 1989 was the founder's great-grandson Walter Lange able to resuscitate A. Lange & Söhne with the help of Mannesmann. Remaining true to the original spirit of his forebearer, he turned the company into an extraordinary *manufacture*. Lange Uhren GmbH meanwhile belongs to the Richemont Group.

Founded:
- in 1845 by Ferdinand Adolph Lange in Glashütte, Germany
- refounded in 1990 by LMH (VDO-Mannesmann)

Headquarters:
Glashütte in Saxony, Germany

Status:
revived (today Richemont Luxury Group)

Specialty:
pocket watches, marine chronometer, flyer and observation watch

Major inventions:
- 1852 club-tooth lever escapement, two-thirds plate
- 1864 three-quarter plate, Glashütte lever escapement of spring-hardened gold

1928
Men's Watch

Case: 14-karat yellow gold, push-down case back, leather strap, 23 x 38 mm
Movement: rhodium-plated, côtes de Genève, 16 jewels, regulated in 5 positions, manual winding
Remarks: one of A. Lange & Söhne's first tonneau-shaped gold wristwatches; regulated in 5 positions; dial signature: "A.Lange & Söhne Genf" (Geneva)
Estimated value: $7,800 →

1932
Men's Watch

Case: 14-karat yellow gold, push-down case back, leather strap, 22 x 38 mm
Movement: gold-plated, frosted finish, manual winding
Remarks: Glashütte men's watch in a rectangular case
Estimated value: $3,250 →

1935
Men's Watch

Case: 14-karat yellow gold, push-down case back, leather strap, Ø 35 mm
Movement: rhodium-plated, côtes de Genève, manual winding
Remarks: rare Lange wristwatch in Bauhaus design
Estimated value: $5,850 →

1935
Men's Watch

Case: 14-karat yellow gold, push-down case back, leather strap, 21 x 36 mm
Movement: Caliber 8.3/4x12''', rhodium-plated, côtes de Genève, manual winding
Remarks: rare men's watch
Estimated value: $3,250 →

1935
Men's Watch

Case: 14-karat yellow gold, push-down case back, leather strap, Ø 36 mm
Movement: 10 1/2 line caliber, nickel-plated, côtes de Genève, manual winding
Remarks: extremely rare men's watch in Bauhaus style with original box and certificate
Estimated value: $7,150 →

1935
Men's Watch

Case: 14-karat yellow gold, push-down case back, leather strap, Ø 36 mm
Movement: 10 1/2-line caliber, rhodium-plated, côtes de Genève, manual winding
Remarks: fine Glashütte men's watch
Estimated value: $5,850 →

1936
Men's Watch

Case: 14-karat yellow gold, push-down case back, leather strap, 21 x 38 mm
Movement: 8 3/4 line caliber, nickel-plated, côtes de Genève, shaped movement, manual winding
Remarks: fine Glashütte men's watch
Estimated value: $3,250 →

1939
Men's Watch with Lange Caliber

Case: 14-karat yellow gold, push-down case back, leather strap, Ø 35 mm
Movement: Caliber 31, gold-plated, frosted finish, jewels set in chatons, manual winding
Remarks: this watch was offered with certification from A. Lange & Söhne's master registry to verify its authenticity
Estimated value: $9,100 →

1936
Observation Watch for the German Air Force

Case: nickel, push-down case back, leather strap, Ø 58 mm
Movement: gold-plated, frosted finish, jewels set in chatons, Glashütte gold pallets, gold escape wheel, manual winding
Remarks: exceptionally rare air force observation watch; according to shipping records, this watch was delivered to the minister for air travel in Berlin on March 31, 1936; only a few pieces of this model were made with optimal rate
Estimated value: $20,800 →

1938
Pilot's Watch with Special Dial

Case: silver, Ø 55 mm
Movement: Caliber 45, gold-plated, frosted finish, Glashütte gold pallets, manual winding
Remarks: extremely rare pilot's observation watch; sold on November 29, 1938 to the minister for air travel and the head of the air force; rare pre-series of the war version of Lange's pilot's watches (Caliber 45 was a small series of 10 pieces)
Estimated value: $23,400 →

1939
Observation Watch for the German Air Force

Case: silver, push-down case back, Ø 55 mm
Movement: gold-plated, frosted finish, Glashütte gold pallets, manual winding
Remarks: exceptionally rare air force observation watch; according to shipping records, this watch was delivered to the minister for air travel in Berlin on September 5, 1939; only nine pieces of this model were made that included a movement with an optimal rate
Estimated value: $32,500 ↗

1940
Observation Watch for the Waffen-SS

Case: silver, push-down case back, leather strap, Ø 65 mm
Movement: 2/3 base plate, gold-plated, frosted finish, jewels set in chatons, compensation watch with stop mechanism, manual winding
Remarks: exceptionally rare observation watch, used for testing batteries of the Waffen-SS. Prototype in silver case with additional seconds display in mirror writing. First watch of this kind for the Waffen-SS
Estimated value: $23,400 →

1940
Pilot's Watch

Case: matte grey, push-down case back, leather strap, Ø 55 mm
Movement: Caliber 48.1, gold-plated, frosted finish, manual winding
Remarks: pilot's watch for the German air force with military engraving
Estimated value: $5,850 →

1940
Pilot's Watch

Case: matte grey, push-down case back, Ø 55 mm
Movement: Caliber 45, gold-plated, frosted finish, Glashütte gold pallets, manual winding
Remarks: exceptionally rare air force observation watch; according to shipping records, this watch was delivered to the minister for air travel in Berlin on January 18, 1940; rare pre-series of the war version of Lange's pilot's watches (Caliber 45 was a small series of 17 pieces)
Estimated value: $13,000 →

1941
Fighter Pilot's Watch for Hermann Goering

Case: silver, push-down case back, leather strap, Ø 55 mm
Movement: gold-plated, frosted finish, manual
Remarks: important, unique silver fighter pilot's observation watch; special order for federal minister of air travel Hermann Goering; sent to Berlin on March 5, 1941; this watch was offered with an excerpt from Lange's master registry and an extra-long pilot's leather strap
Estimated value: $52,000 →

1943
Pilot's Watch

Case: painted, push-down case back, leather strap, Ø 55 mm
Movement: gold-plated, frosted finish, manual winding
Remarks: pilot's watch for the German air force
Estimated value: $5,850 →

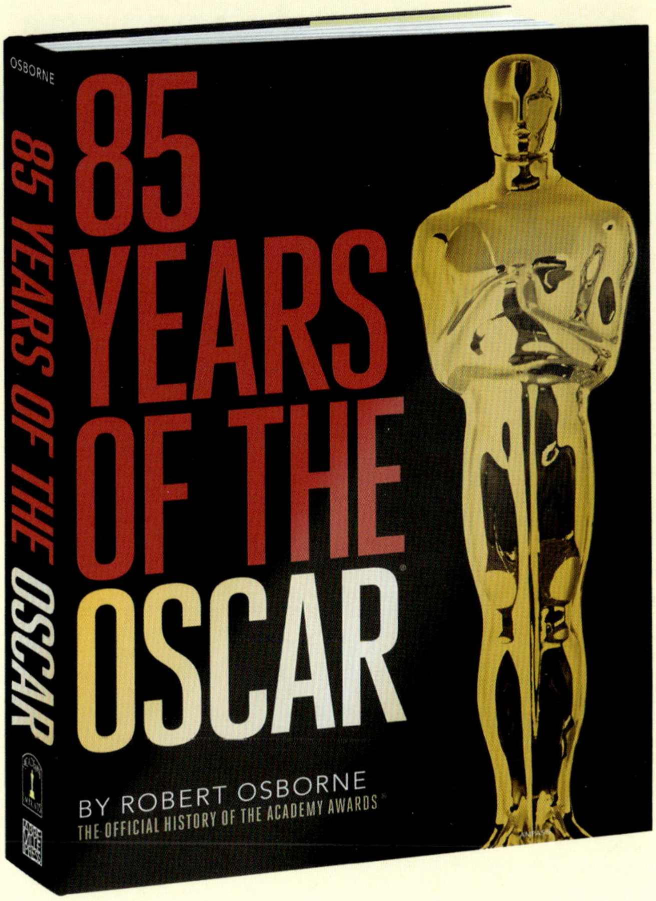

85 Years of the Oscar®

The Official History of the Academy Awards®

The only official history of the Academy Awards, *85 Years of the Oscar* captures the thrill of the film industry's most significant and popular event. Written by film historian and Hollywood insider Robert Osborne, *85 Years of the Oscar*, with fascinating background information and more than 750 photographs, provides an authority and depth of coverage found nowhere else.

Detailing the history of the Awards from the formative years of the film industry to the present, Osborne shows the influence of changing world events on the industry, details iconic moments from each year's ceremony, and reveals how the movies honored each year are a reflection of their time. *85 Years of the Oscar* is meticulously researched and written in association with the Academy.

By Robert Osborne
765 illustrations, 96 in full color
472 pages · 9 × 12"
Cloth · $75.00
ISBN 978-0-7892-1142-2

"Robert Osborne keeps the legend of Hollywood alive and well!"—*Drew Barrymore*

"I say this softly, the Oscar along with the Academy or the Academy along with the Oscar, could not have a better film historian or human being who loves what the Academy and statue stands for."
—*A friend, Bill Cosby*

"No one knows more about Oscar history—or chronicles it with greater enthusiasm—than the one and only Robert Osborne. I look forward to each new edition of this handsome and valuable book." —*Leonard Maltin*

"A must own for any serious fan of film."
—*IMDb.com*, The Internet Movie Database

Published by ABBEVILLE PRESS
137 Varick Street, New York, NY 10013
1-800-ARTBOOK (in U.S. only)
Also available wherever fine books are sold
Visit us at www.abbeville.com

ANGELUS

It was the Stolz brothers Gustave and Albert who founded Angelus in Le Locle in 1891, and the company quickly established a fine reputation for chronographs and repeaters. The quality and creativity of its products were recognized at the Swiss national exhibition in 1914, when the company was given an award for its pocket watches and repeater movements.

The Stolz brothers began manufacturing wristwatches early on, especially chronographs, and today the name Angelus continues to be associated primarily with these stopwatches. The company developed its own calibers, notably the Caliber 215, for use in complicated models such as the Chronodato.

Alongside the classic Chronodato featuring a date hand and displays of the day and the month in two separate windows, Angelus introduced another variation on the theme in 1942: a model featuring a digital date display. Since the watchmakers used two date disks to achieve this, they could fit a relatively large display into a small amount of space. The Chronodato with digital date display can well be viewed as one of the predecessors of today's large-date chronographs, and it was a complete success even back then.

Angelus also manufactured standard three-handed watches a well as timepieces that were often outfitted with a calendar function, a moon phase display, or a large date.

In 1957, Angelus conceived a very special watch called the Tinkler. This English term was applied to describe the timepiece's specialty: the automatic watch was equipped with a repeater movement. The watchmakers chose AS Caliber 1580 as their base, which they proceeded to outfit with a modular repeater mechanism—on the dial side, since the back was already taken up with the automatic winding mechanism. However, Angelus's supplier, AS in Grenchen, demanded a production volume of at least 10,000 pieces, and since Angelus was not prepared to deal with such a large quantity at one time, the Tinkler remained a small series of 100 pieces encased in steel—watches that are still rarities to this day.

Even if the company's experience with the Tinkler was not the best, Angelus tried to realize the true potential of the concept once again in 1978. The base this time was an ETA caliber for which Dubois Dépraz created a five-minute repeater module, produced by Kelek. At the time, however, Angelus did not have the financial means to successfully introduce this large watch on the market. Since mechanical movements were no longer in demand in the 1970s, and the company had failed to get a jump on quartz watches in a timely manner, it wasn't long before Angelus fell victim to the quartz crisis. Today, all that is left of this manufacturer of excellent chronographs is a large sign featuring the Angelus logo on the old factory in Le Locle.

Founded:
1891 by Albert and Gustave Stolz

Headquarters:
Le Locle

Status:
extinguished, company liquidated

Specialty:
chronographs, calendar watches, repeaters, table alarm clocks

Major inventions:
1949 large date

Leading models:
Chronodato

1945
Chronograph
Case: 18-karat yellow gold, push-down case back, leather strap, Ø 35 mm
Movement: nickel-plated, column-wheel control of chronograph, manual winding
Remarks: chronograph with 45-minute counter
Estimated value: $1,040 →

1945
Chronograph
Case: 18-karat yellow gold, push-down case back, leather strap, Ø 35 mm
Movement: nickel-plated, côtes de Genève decoration, column-wheel control of chronograph, manual winding
Remarks: chronograph with 45-minute counter
Estimated value: $1,040 →

1945
Chronodato
Case: stainless steel; Ø 38 mm; tripartite, stainless steel push-down case back, leather strap
Movement: rhodium-plated, decorated, finely polished, beveled chronograph steel parts, column-wheel control of chronograph
Remarks: chronograph with 45-minute counter; complete calendar with date hand and window display of day, date, and month
Estimated value: $1,560 →

1945
Chronodato
Case: stainless steel, push-down case back, leather strap, Ø 38 mm
Movement: rhodium-plated, côtes de Genève decoration, manual winding
Remarks: chronograph with 45-minute counter; complete calendar; date ring different color from dial
Estimated value: $1,820 →

1950
Chronograph
Case: stainless steel, screw-down case back, leather strap, Ø 38 mm
Movement: rhodium-plated, column-wheel control of chronograph, manual winding
Remarks: chronograph with 45-minute counter
Estimated value: $910 →

1950
Chronodato
Case: gold-plated stainless steel, push-down case back, leather strap, Ø 38 mm
Movement: nickel-plated, column-wheel control of chronograph, manual winding
Remarks: chronograph with 45-minute counter; complete calendar
Estimated value: $1,300 →

1958
Tinkler
Case: 18-karat yellow gold, screw-down case back, leather strap, Ø 36 mm
Movement: Caliber AS 1580, rhodium-plated, automatic
Remarks: very rare men's watch with quarter repeater module from a prototype series of 100 watches; a button at 9 o'clock activates the repeater
Estimated value: $3,900 →

1960
Date Alarm
Case: stainless steel, screw-down case back, leather strap, Ø 34 mm
Movement: Caliber AS 1568N, nickel-plated, twin spring barrels, 21 jewels, manual winding
Remarks: alarm wristwatch with AS Caliber 1568
Estimated value: $390 →

AUDEMARS PIGUET

In Le Brassus in 1875, the youngest of the Grenoble watchmaker family Audemars, Jules-Louis, founded his own workshop in his family's house, where he planned to make complicated movements of the best quality.

Just a short while later, Audemars had actually gotten somewhere: he was closing good contracts and his watches were in demand in Geneva. In order to avoid delivery problems, he employed various watchmakers from his family and circle of friends in his workshop—among them Edward-Auguste Piguet, someone he knew from school and choir.

Since Piguet was also dedicated to high-quality watches, the pair decided to establish a joint-stock company in 1881: Audemars Piguet SA. From that point on, they stopped making calibers for other companies. They manufactured pocket watches under their own name, most of which were outfitted with additional functions such as repeater mechanisms. Thanks to their success, they were able to open a subsidiary in Geneva in 1889. The brand's headquarters would remain in Le Brassus all the way to the modern day, however.

The pair used the 1889 World's Fair in Paris to introduce a *grande complication* that attracted quite a bit of attention: this pocket watch contained a minute repeater, a chronograph, and a perpetual calendar.

Their small workshop was no longer large enough for them, and they built a new factory in Le Brassus in 1907 that is still in use today, though it has been expanded a few times.

In 1918, Audemars passed away, and his partner Piguet followed him one year later. Their sons Paul-Louis Audemars and Paul-Edward Piguet took over the company. Ten years later, the Depression hit along with hard times for the company; Audemars Piguet was driven to the brink of bankruptcy. The two owners could only save themselves by premiering more reasonably priced watches and going back to the company's roots by once again making movements for other brands.

Audemars Piguet continued to justify its reputation with its wristwatches: alongside high-quality three-handed watches, it was the calendar models that were the most popular. And the company's reputation would not have been justifiable at all if there had been no perpetual calendar in its program. Jumping digital hour displays and attractive skeletonized watches also belonged to Audemars Piguet's repertoire.

The movements in Audemars Piguet's watches had always been of excellent quality: in 1946, the company introduced a manually wound movement only 1.46 mm in height that continues to be manufactured to this day. The same is true of the slimmest automatic movement of its day, presented in 1967: its height of 2.45 mm including a rotor on jewel bearings.

The company became absolutely legendary in 1972 with an extremely modern stainless steel watch that accepted no compromise: this Audemars Piguet version of a sports watch was called the Royal Oak, and it featured an octagonal bezel. And even though it was a steel watch, it cost just as much as some other brands' gold models. Its shape and price confused not only customers at first—within the company there were also varying opinions on the new model. Despite these early growing pains, the Royal Oak went on to become the company's flagship model.

Audemars Piguet is one of the few brands still owned by its founding family uninterruptedly since its establishment—now for more than 130 years.

Founded:
1875 by Jules-Louis Audemars

Headquarters:
Le Brassus

Status:
active without interruption

Specialty:
minute repeaters, tourbillon, chronographs, ultra-thin watches

Major inventions:
- 1892 first wristwatch with a minute repeater
- 1972 Royal Oak
- 1987 first wristwatch with a tourbillon and automatic winding

Leading models:
Royal Oak

1925
Jump Hour

Case: 18-karat white gold, push-down case back, leather strap, 25 x 37 mm
Movement: rhodium-plated, côtes de Genève decoration, 18 jewels, regulated in 8 positions, manual winding
Remarks: extremely rare men's watch with a digital jump hour display, regulated in 8 positions
Estimated value: $10,400 →

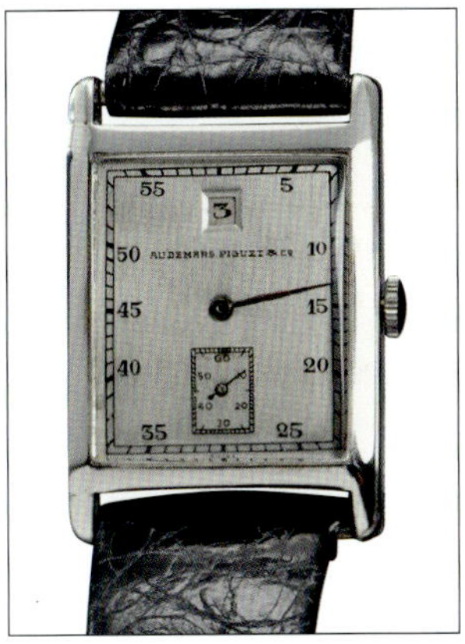

1926
Men's Watch

Case: 18-karat white gold, push-down case back with monogram, white gold bracelet, 23 x 39 mm
Movement: rhodium-plated bridges, fausses côtes decoration, compensation screw balance, 17 jewels, regulated in 5 positions, manual winding
Remarks: heavy, very fine man's wristwatch
Estimated value: $5,200 →

1929
Men's Watch with Digital Display

Case: 18-karat white gold, push-down case back, leather strap, 27 x 31 mm
Movement: Caliber GHSM, rhodium-plated, côtes de Genève decoration, 17 jewels, regulated in 3 positions, manual winding
Remarks: significant early men's watch with digital display, regulated in 3 positions
Estimated value: $13,000 →

1943
Chronograph with Pulsometer

Case: platinum, tripartite, push-down case back, leather strap, 38 x 44 mm
Movement: rhodium-plated, fausses côtes decoration, fine matte steel chronograph components, column wheel control of chronograph
Remarks: elegant doctor's watch in a platinum case with a pulsometer scale on the enamel dial; the chronograph button is integrated into the crown
Estimated value: $52,000 ↗

1946
Men's Watch

Case: platinum, push-down case back, leather strap, 25 x 33 mm
Movement: Caliber 9/10RS, nickel-plated, côtes de Genève decoration, 18 jewels, regulated in 8 positions, manual winding
Remarks: rare platinum men's watch with movement regulated in 8 positions; the watch was offered in its original box and with a certificate from Audemars Piguet
Estimated value: $3,900 →

1948
Men's Watch

Case: 18-karat yellow gold, leather strap, 26 x 36 mm
Movement: nickel-plated, côtes de Genève decoration, 18 jewels, regulated in 8 positions, manual winding
Remarks: rare men's watch with rectangular case
Estimated value: $2,340 ↗

1949
Chronograph

Case: 18-karat yellow gold, push-down case back, leather strap, Ø 33 mm
Movement: rhodium-plated, côtes de Genève decoration, column-wheel control of chronograph, 21 jewels, regulated in 8 positions, manual winding
Remarks: chronograph with 30-minute counter; this watch has a dedication for a later president of the Coca Cola Corporation engraved on the case back; offered in its original box and with its original certificate
Estimated value: $26,000 ↗

1942
Chronograph

Case: 18-karat yellow gold, two-part, push-down case back, leather strap, Ø 33 mm
Movement: rhodium-plated, fausses côtes decoration, finely polished, beveled steel chronograph parts, column wheel control, glucydur screw balance, manual winding
Remarks: rare men's chronograph with a very fine movement
Estimated value: $32,500 ↗

1948
Men's Watch

Case: 18-karat yellow gold, push-down case back, leather strap, Ø 35 mm
Movement: rhodium-plated, fausses côtes decoration, glucydur screw balance, indirect sweep second, 18 jewels, regulated in 8 positions, manual winding
Remarks: lateral pusher to stop balance for time-setting to the second following a time signal
Estimated value: $5,850 →

1955
Men's Watch for Cartier

Case: 18-karat yellow gold, push-down case back, leather strap, Ø 32 mm
Movement: Caliber 2001LEC, rhodium-plated, côtes de Genève decoration, manual winding
Remarks: men's watch signed by both Audemars Piguet and Cartier; made for Cartier
Estimated value: $5,850 →

1957
Men's Watch

Case: 18-karat yellow gold, push-down case back, leather strap, Ø 35 mm
Movement: Caliber V 7SSC, nickel-plated, côtes de Genève decoration, regulated in 5 positions, manual winding
Remarks: men's watch with movement regulated in 5 positions; this watch was offered in its original box
Estimated value: $2,600 →

1956
Automatic

Case: 18-karat yellow gold, bipartite, screw-down case back, leather strap, Ø 35 mm
Movement: Caliber P2498, rhodium-plated, fausses côtes decoration, gold rotor, polished screws, 21 jewels, regulated in 5 positions
Remarks: classic automatic men's watch with subsidiary seconds; the movement has a gold rotor
Estimated value: $4,550 ↗

1961
Automatic

Case: 18-karat yellow gold, screw-down case back, leather strap, Ø 35 mm
Movement: Caliber 2072LEC, rhodium-plated, decorated, gold rotor in jewel bearings, 29 jewels, regulated in 5 positions, automatic
Remarks: automatic men's watch with date; this watch was offered in a box
Estimated value: $5,850 ↗

1969
Automatic

Case: 18-karat yellow gold, tripartite, push-down case back, fluted case band, leather strap
Movement: Caliber 2071 LEC, rhodium-plated, fausses côtes decoration, 18-karat gold rotor in jewel bearings, polished screws, 29 jewels, regulated in 5 positions
Remarks: elegant men's watch with 18-karat gold rotor on jewel bearings
Estimated value: $4,550 →

1965
Automatic

Case: 18-karat white gold, screw-down case back, leather strap, Ø 35 mm
Movement: Caliber K2072, nickel-plated, côtes de Genève decoration, 18-karat gold rotor on jewel bearings, automatic
Remarks: automatic men's watch with gold buckle; this watch was offered in its original box
Estimated value: $5,850 →

1972
Automatic

Case: 18-karat white gold, tripartite, leather strap, Ø 32 mm
Movement: Caliber K2120, rhodium-plated, fausses côtes decoration, gold rotor, Gyromax balance, 36 jewels, regulated in 6 positions
Remarks: small men's watch in a white gold case
Estimated value: $3,900 ↗

1972
Men's Watch

Case: 18-karat white gold, push-down case back, leather strap, Ø 33 mm
Movement: rhodium-plated, côtes de Genève decoration, 18 jewels, regulated in 8 positions, manual winding
Remarks: elegant men's watch with movement regulated in 8 positions
Estimated value: $2,340 →

1972
Royal Oak

Case: stainless steel, monocoque case, stainless steel link bracelet, Ø 39 mm, water-resistant to 100 m
Movement: Caliber 2121, rhodium-plated, automatic
Remarks: first luxury sports watch in stainless steel; early model still with monocoque case; with a retail price of 3,650 Swiss francs in 1972, this was a very expensive watch in its day
Estimated value: $15,600 ↗

1974
Men's Watch

Case: 18-karat white gold, push-down case back, leather strap, 33 x 34 mm
Movement: Caliber 2003/1, nickel-plated, côtes de Genève decoration, regulated in 8 positions, manual winding
Remarks: rare men's watch with diamond-set hour markers and hands; movement regulated in 8 positions
Estimated value: $3,250 ↗

1974
Automatic

Reference number: 5420
Case: 18-karat white gold, push-down case back, leather strap, 32 x 36 mm
Movement: Caliber K2121, rhodium-plated, côtes de Genève decoration, regulated in 5 positions, automatic
Remarks: men's watch with movement regulated in 5 positions and a 21-karat gold rotor
Estimated value: $3,250 ↗

1972
Skeleton Watch

Case: 18-karat yellow gold, push-down transparent case back, leather strap, Ø 33 mm
Movement: Caliber 2120, gold rotor, movement completely skeletonized and decorated by hand, glucydur balance, 36 jewels, regulated in 8 positions, automatic winding
Remarks: very fine skeleton wristwatch, small series production
Estimated value: $5,200 →

1984
Perpetual Calendar Automatic

Case: 18-karat white gold, diamond bezel, push-down case back, white gold bracelet, Ø 38 mm
Movement: Caliber 2120/2, rhodium-plated gold rotor, fausses côtes decoration, glucydur balance, 38 jewels, regulated in 8 positions, automatic winding
Remarks: perpetual calendar and moon phase display; dress watch produced in a limited small series
Estimated value: $32,500 →

1986
Tourbillon Automatique

Case: 18-karat yellow gold, diamond bezel, push-down case back, leather strap, 32 x 28 mm
Movement: 1-minute tourbillon, finely matted titanium cage, 32 jewels, automatic winding with eccentric oscillating mass, crown at case back
Remarks: world's first (1986) and for a long time thinnest wristwatch tourbillon with automatic winding
Estimated value: $13,000 ↗

1987
Chronograph Automatic

Case: 18-karat yellow gold, push-down case back, yellow gold bracelet, Ø 40 mm
Movement: Caliber 2126, rhodium-plated gold rotor, fausses côtes decoration, glucydur balance, 51 jewels, regulated in 8 positions, automatic winding
Remarks: heavy gold chronograph with automatic winding
Estimated value: $13,000 →

BAUME & MERCIER

This brand's two namesakes, Louis-Victor Baume and Paul Mercier, actually lived a century apart from each other. Louis-Victor Baume (1817–1887) was a watchmaker hailing from the forests north of La Chaux-de-Fonds. Together with his brother Pierre Joseph Célestin, he registered a company called Les Frères Baume in 1830 and invited their younger brothers to join the workshop a few years later. The small company experienced a quick upward swing, especially in England, where in 1851 a flourishing subsidiary was founded under the name Baume Brothers, which was to later become completely independent.
One of the Baume brothers' specialties was precise technical watches: a chronograph with a tourbillon, signed Baume & Co., set a new record in precision at a chronometry competition organized by the royal observatory in Kiev. The Baume firm also took part in all the World's Fairs between 1860 and 1910, wining numerous gold medals with its watches.
William-Adolphe Baume, born in 1885, Louis-Victor's grandson, had already been working in the family business since the days of his youth when, in 1912, he met watchmaker and jeweler Paul Mercier, whose birth name was actually Cherednichenko. His Russian father had met a charming seamstress in Paris and fell in love with her. After the father died in an accident in Russia, Mercier, his mother, and his two sisters moved to Switzerland, where he found work with Genevan chronometer maker Haas. As a creative individual and someone who knew how to enjoy life, Mercier took over Baume & Co.'s commercial management with much success. The serious and rather taciturn William Baume remained in charge of technology. Together, they established the brand Baume & Mercier in 1920. The *manufacture* continued to make precision watches, and in the first year of its new existence, it enjoyed a reputation as the Genevan chronometer office's best customer. Trendy wristwatches and reliable chronographs were Baume & Mercier's main sources of turnover.
The founders left the company in the mid-1930s, and new owners Ponti and Gennari, jewelers from Turin, installed Constantin de Gorski as managing director, who in turn thoroughly modernized both the collection and the production equipment. After World War II, Baume & Mercier made a reputation for itself with innovative ladies' watches such as the Marquise model featuring a spring-action bangle (at right) and gained in international prestige.
In 1952, Baume & Mercier took over C.H. Meylan Watch SA in Le Brassus, famous for ultra-flat movements, thus achieving additional autonomy. After the death of de Gorski, the brand lost momentum, and the Piaget family, with which the company had collaborated for quite some time, took over the majority in 1964. The two brands blossomed in the shape-obsessed 1970s with avant-garde and innovative timepieces, some outfitted with ultra-flat automatic movements or quartz calibers, before they were purchased by Cartier in 1988, which was later integrated into the Vendôme Group and, finally, Richemont.

Founded:
- 1830 by Louis-Victor and Pierre Joseph Céléstin Baume in Le Bois
- 1918 new founding by William Baume in Geneva
- 1920 renamed Baume & Mercier

Headquarters:
Geneva

Status:
active without interruption (today Richemont Luxury Group)

Specialty:
extra-thin and ultra-thin ladies' and men's watches, chronographs, ladies' jewelry watches

1942
Ladies' Watch

Case: 14-karat yellow gold, push-down case back, leather strap, 34 mm
Movement: Caliber AS 1234, rhodium-plated, 17 jewels, manual winding
Remarks: ladies' watch in shaped gold case
Estimated value: $325 →

archive photo

1945
Chronograph

Case: 18-karat rose gold, monocoque case, leather strap
Movement: Caliber W69, rhodium-plated, Breguet hairspring, 17 jewels, manual winding
Remarks: chronograph with 30-minute counter and tachymeter scale
Estimated value: $3,250 →

archive photo

1945
Ladies' Watch

Case: stainless steel, push-down case back, gold strap lugs, leather strap
Movement: Caliber AS 1234, rhodium-plated, 17 jewels, manual winding
Remarks: rare ladies' watch in a bipartite, two-tone case
Estimated value: $185 →

archive photo

1945
Calendar Watch

Case: stainless steel, push-down case back, leather strap, 24 x 37 mm
Movement: Caliber ETA 1164, rhodium-plated, 17 jewels, manual winding
Remarks: rare calendar watch (date hand, window display for days and months); corrector on case back
Estimated value: $780 →

archive photo

1945
Men's Watch

Case: 18-karat yellow gold, monocoque case, leather strap
Movement: Caliber Felsa 690, rhodium-plated, 17 jewels, automatic
Remarks: elegant automatic watch with sweep seconds
Estimated value: $910 →

archive photo

1948
Men's Watch

Case: 18-karat rose gold, push-down case back, leather strap
Movement: Caliber BM 960, rhodium-plated, 15 jewels, manual winding
Remarks: elegant men's watch with subsidiary seconds
Estimated value: $520 →

archive photo

1953
Calendar Chronograph

Case: 18-karat rose gold, push-down case back, leather strap
Movement: Caliber W72, rhodium-plated, Breguet hairspring, 17 jewels, manual winding
Remarks: chronograph with 30-minute and 12-hour counters; complete calendar with date hand
Estimated value: $1,970 →

archive photo

1954
Chronograph

Case: 18-karat rose gold, push-down case back, leather strap
Movement: Landeron Caliber 48, nickel-plated, 17 jewels, manual winding
Remarks: chronograph with 30-minute counter and tachymeter scale
Estimated value: $1,300 →

archive photo

BREITLING

Léon Breitling, a German watchmaker originally, founded his firm in Saint-Imier in 1884. He began by building measurement instruments for the production of watches, but he soon began manufacturing stopwatches. In 1892, Breitling had already moved his headquarters to La Chaux-de-Fonds, changed the business name to Léon G. Breitling, Montbrillant Watch Manufactory, and started making chronographs. The watches were famous for their reliability, proving their mettle on the battlefields of World War I.

Léon's son Gaston, and later his grandson Willy, made this company into one of the leading chronograph manufacturers based on this reputation. The Breitling name has always stood for quality, robustness, and reliability. Thanks to this reputation, it was Breitling's watches that were the first choice in outfitting the cockpits of the 1930s with the best chronographs. These watches, indispensable for navigation, had to be both precise and robust enough to survive the increasing acceleration powers of modern aviation. Although Breitling didn't manufacture its own movements, the company's assembly specialists were able to tailor the watches perfectly to the needs of the pilot.

The company introduced a singular highlight in 1952, which would from then on be a status symbol and object of desire for pilots all over the world: the Navitimer.

What was so special about this timepiece? A slide rule was built into the bezel, making the watch a universal navigation instrument that could take over the calculation of flight time and energy consumption long before computers. This was also the watch to travel with the first expedition to space on the wrist of astronaut Scott Carpenter on board the *Mercury* capsule.

In the late 1960s, Breitling, together with Heuer, Hamilton-Büren, and Dubois Dépraz, heralded the development of the automatic chronograph. After a few years of research, the four companies were able to present the fruit of their common labors in 1969: Caliber 11. Together with Calibers 12 and 15, which came later, this was long the only alternative to Zenith's El Primero with regard to automatic chronographs.

Despite this exceptional innovation, and even though Breitling had gotten an early start becoming familiar with quartz watches in a changing world market, the company made a crash landing in 1979, going bankrupt. It was bought the same year by Ernest Schneider, a businessman and enthusiastic pilot. At his bidding, the company moved from La Chaux-de-Fonds to Grenchen.

Schneider has followed a consistent strategy: although a much less expensive quartz watch may measure the time more precisely, the brand never turned its back on mechanical chronographs.

Founded:
- 1884 by Léon Breitling in St.-Imier
- 1982 new founding

Headquarters:
Grenchen since 1982

Status:
active

Specialty:
chronograph and pilot's watches

Major inventions:
- 1923 separate chronograph pushers
- 1934 zero-reset pusher
- 1942 integrated slide rule
- 1969 co-development of the automatic chronograph Caliber 11
- 1995 Emergency with flight emergency call system

Leading models:
Navitimer, Chronomat, Chrono-Matic

1945
Chronograph
Reference number: 765
Case: stainless steel, tripartite, push-down case back, Ø 37 mm
Movement: rhodium-plated, fine matte steel chronograph components, column wheel control of chronograph
Remarks: chronograph with 30-minute and 12-hour counters
Estimated value: $1,820 →

1945
Chronograph Premier
Reference number: 788
Case: stainless steel, tripartite, screw-down case back, leather strap, Ø 35 mm
Movement: rhodium-plated, fine matte steel chronograph components, column wheel control of chronograph
Remarks: chronograph with 30-minute and 12-hour counters, tachymeter scale
Estimated value: $2,210 →

ca. 1945
Chronograph
Reference number: 801
Case: stainless steel case, tripartite with push-down case back, rotating bezel, leather strap, Ø 36 mm
Movement: manual winding, 17 jewels, rhodium-plated, column wheel control of chronograph
Remarks: extremely rare pilot's chronograph with 30-minute and 12-hour counter, date and moon phase at 12, integrated slide rule and hundredth divisions
Estimated value: $10,400 ↗

1945
Chronograph Premier
Case: steel case with push-down case back, Ø 37 mm
Movement: manual winding
Remarks: simple manually wound chronograph with 45-minute counter; the 3, 6, and 9 markers are especially accentuated for special legibility of stopped time for long-distance phone calls, which were calculated in three-minute intervals
Estimated value: $1,300 →

1945
Chronograph Premier
Reference number: 788
Case: 18-karat yellow gold, screw-down case back, faceted, Ø 36 mm
Movement: manual winding
Remarks: simple manually wound chronograph with 30-minute and 12-hour counters, the 3, 6, and 9 markers are especially accentuated for special legibility of stopped time for long-distance phone calls, which were calculated in 3-minute intervals
Estimated value: $4,550 →

1945
Chronograph Premier
Reference number: 777
Case: 18-karat gold, screw-down case back, faceted, Ø 37 mm
Movement: manual winding
Remarks: simple manually wound chronograph with 45-minute counter
Estimated value: $3,900 ↗

1945
Chronomat
Case: stainless steel, push-down case back, rotating bezel, faceted, Ø 36 mm
Movement: manual winding
Remarks: manually wound chronograph with 45-minute counter, additional hundredth scale for the chronograph seconds
Estimated value: $3,250 →

1950
Calendar Watch
Case: 18-karat yellow gold, push down case back, faceted, Ø 36 mm
Movement: Caliber 20, manual winding
Remarks: rare, simple watch with date hand and display of weekday and month in two small windows; subsidiary seconds at 6 o'clock
Estimated value: $2,340 →

1950
Men's Watch

Case: gold-plated stainless steel, push-down case back, Ø 33 mm
Movement: Felsa Caliber 4010, manual winding
Remarks: simple three-handed watch
Estimated value: $260 →

1950
Split-Seconds Chronograph

Reference number: 762
Case: steel case with push-down case back, faceted, Ø 36 mm
Movement: manual winding
Remarks: split-seconds manually wound chronograph with 45-minute counter, the split-seconds function makes it possible to take intermediate timing without interrupting the ongoing timing
Estimated value: $10,400 ↗

1949
Chronograph Datora

Case: gold-plated case with steel push-down case back, Ø 35 mm
Movement: manual winding
Remarks: calendar chronograph with 30-minute and 12-hour counters; additional date and moon phase displays at 12 o'clock
Estimated value: $4,550 →

1960
Chronomat

Reference number: 808
Case: stainless steel, push-down case back, rotating bezel, Ø 37 mm
Movement: Venus Caliber 175, nickel-plated, column wheel control of chronograph, manual winding
Remarks: manually wound chronograph with multifunctional bezel with slide rule function, 45-minute counter, Alpha steel hands; this watch has box and papers
Estimated value: $3,380 →

1965
Split-Seconds Chronograph

Case: 18-karat yellow gold, push-down case back, Ø 36 mm
Movement: manual winding
Remarks: rare manually wound chronograph with 30-minute counter and split-seconds function; the split-seconds function makes it possible to take intermediate timing without interrupting the ongoing timing
Estimated value: $11,700 ↗

1960
Chronograph AVI

Reference number: 765
Case: steel case, tripartite, screw-down case back, rotating bezel, Ø 41 mm
Movement: Venus Caliber 178, column wheel control of chronograph, rhodium-plated, manual winding
Remarks: manually wound chronograph; extremely rare variant with digital 15-minute and 12-hour counters; fine matte steel components
Estimated value: $6,500 ↗

1969
Navitimer

Reference number: 806
Case: stainless steel, push-down case back, rotating bezel, Ø 40 mm
Movement: Venus Caliber 178, gold-plated, column wheel chronograph control, manual
Remarks: manually wound chronograph with 30-minute and 12-hour counters; column wheel control of chronograph, glucydur balance, bezel can be used as a slide rule; this watch was offered with original box and papers
Estimated value: $3,380 →

1973
Chronomat

Reference number: 7808
Case: stainless steel, tripartite, push-down case back, rotating bezel, Ø 40 mm
Movement: Valjoux Caliber 7740, rhodium-plated, fine matte steel chronograph components
Remarks: chronograph with 30-minute and 12-hour counters; rotating bezel can be used as slide rule
Estimated value: $3,640 →

1973
Navitimer

Reference number: 7806
Case: stainless steel, push-down case back, rotating bezel, Ø 40 mm
Movement: Valjoux Caliber 7740, rhodium-plated, fine matte steel components, manual winding
Remarks: manually wound chronograph with 30-minute and 12-hour counters; date window between 4 and 5 o'clock; rotating bezel can be used as a slide rule
Estimated value: $3,250 →

1975
Chronomat Chrono-Matic

Reference number: 8808.3
Case: stainless steel, push-down case back, rotating bezel, Ø 41 mm
Movement: Caliber 12, gold-plated, fine matte steel components, eccentric fine adjustment, automatic
Remarks: automatic chronograph with 30-minute counter; date at 6 o'clock; rotating bezel can be used as a slide rule
Estimated value: $3,640 →

1975
Cosmonaute

Reference number: 809.4
Case: gold-plated, stainless steel push-down case back, rotating bezel, Ø 40 mm
Movement: Venus Caliber 178, gold-plated, column wheel control of chronograph, manual
Remarks: manually wound chronograph with 24-hour dial; 30-minute and 12-hour counters; glucydur balance; bezel was conceived to be used as a slide rule; this watch was offered with original box and papers
Estimated value: $3,640 →

1975
Cosmonaute

Reference number: 809
Case: stainless steel, screw-down case back, rotating bezel, Ø 41 mm
Movement: Venus Caliber 178, gold-plated, column wheel control of chronograph, manual winding
Remarks: manually wound chronograph with 24-hour dial; 30-minute and 12-hour counters; rotating bezel can be used as a slide rule
Estimated value: $3,640 →

1973
Chronomat

Reference number: 7808
Case: stainless steel, tripartite, push-down case back, rotating bezel, Ø 41 mm
Movement: Valjoux Caliber 7740, rhodium plated, fine matte steel chronograph components
Remarks: large chronograph with 30-minute and 12-hour counters; date window display; rotating bezel can be used as slide rule
Estimated value: $3,380 →

1963
Chronomat

Reference number: 2110
Case: stainless steel, screw-down case back, rotating bezel, Ø 40 mm
Movement: Caliber 11, nickel-plated, fine matte steel components, automatic
Remarks: automatic chronograph with 30-minute and 12-hour counters; 30-minute scale divided into five blocks for better legibility; additional tachymeter scale
Estimated value: $1,560 →

1968
Super Ocean Regatta

Reference number: 7652
Case: stainless steel, tripartite, screw-down case back, rotating bezel, Ø 48 mm
Movement: Venus Caliber 178, red gold-plated, fine matte steel chronograph components, column wheel control of chronograph
Remarks: large chronograph with 15-minute and 12-hour counters; the watch is outfitted wtih a double regatta time measurement especially for sailors
Estimated value: $3,900 →

1968
Navitimer

Reference number: 1860
Case: stainless steel, screw-down case back, rotating bezel, Ø 48 mm
Movement: Caliber 12, fine matte steel components, micro rotor, eccentric fine adjustment, automatic
Remarks: automatic chronograph with 30-minute and 12-hour counter, date at 6 o'clock
Estimated value: $3,900→

1975
Cosmonaute Super Ocean

Reference number: 2105
Case: stainless steel, screw-down case back, rotating bezel, Ø 49 mm
Movement: Caliber 12, rhodium-plated, automatic
Remarks: automatic chronograph with 15-minute and 6-hour counters; date window at 6 o'clock; unusual marker intervals are highly noticeable along the case sides
Estimated value: $3,380 →

1977
Navitimer Automatic

Reference number: 1806
Case: stainless steel, tripartite, screw-down case back, rotating bezel, leather strap, Ø 48 mm
Movement: Caliber 12, gold-plated, fine matte chronograph stainless steel components, micro rotor, fine adjustment
Remarks: large navigational chronograph for the Iraqi Air Force with 30-minute and 12-hour counters, date window at 6 o'clock, slide rule function above rotating bezel
Estimated value: $4,550 ↗

1977
Chrono-Matic

Reference number: 2130
Case: stainless steel, screw-down case back, rotating bezel, Ø 38 mm
Movement: Caliber 12, gold-plated, fine matte steel components, automatic
Remarks: automatic chronograph with 30-minute counter; polished steel components; glucydur balance
Estimated value: $1,950 →

1967
Co-Pilot

Reference number: 765 CP
Case: stainless steel, tripartite, screw-down case back, rotating bezel, leather strap, Ø 40 mm
Movement: Venus Caliber 178, red gold-plated, fine matte steel chronograph components, column wheel control of chronograph, 17 jewels
Remarks: rare pilot's chronograph with 15-minute and 12-hour counters; a second hour scale is printed onto the rotating bezel
Estimated value: $5,200 ↗

1969
Chronograph Top Time

Reference number: 810
Case: stainless steel, push-down case back, Ø 38 mm
Movement: Venus Caliber 178 TJ, red gold-plated, column wheel control of chronograph, manual winding
Remarks: manually wound chronograph with 30-minute and 12-hour counters; this watch was offered with original box and papers
Estimated value: $2,600 →

1960
Chronograph Cadette

Reference number: 1155
Case: steel case, tripartite, push-down case back, Ø 34 mm
Movement: Valjoux Caliber 7733, rhodium-plated, manual winding
Remarks: manually wound chronograph with 30-minute counter; polished steel chronograph components; monometallic balance; Incabloc shock protection
Estimated value: $490 →

1969
Chronograph Top Time

Reference number: 815.4
Case: gold-plated, push-down case back, Ø 38 mm
Movement: Valjoux Caliber 7736 TJ, manual winding
Remarks: manually wound chronograph with 30-minute and 12-hour counters; polished steel parts; monometallic balance
Estimated value: $1,950 →

1969
Chronograph Top Time

Case: gold-plated, tripartite, steel push-down case back, Ø 35 mm
Movement: manual winding
Remarks: simple manually wound chronograph; the chronograph second and minute hands are highlighted in orange for better legibility
Estimated value: $1,300 →

1969
Chrono-Matic

Reference number: 2112
Case: stainless steel, screw-down case back, rotating bezel, Ø 39 mm
Movement: Caliber 11, nickel-plated, fine matte steel components, automatic
Remarks: automatic chronograph with 30-minute and 12-hour counters; 30-minute scale divided into five blocks for better legibility; additional tachymeter scale
Estimated value: $1,300 →

1970
Chronograph Long Playing

Reference number: 7103.3
Case: stainless steel, push-down case back, 40 x 47 mm
Movement: Valjoux Caliber 7740, nickel-plated, manual winding
Remarks: manually wound chronograph with 30-minute and 12-hour counters; date window between 4 and 5 o'clock
Estimated value: $1,820 ↗

1975
Chronograph Football

Reference number: 2734.3
Case: stainless steel, push-down case back, rotating bezel, 41 x 47 mm
Movement: Valjoux Caliber 7731, manual winding
Remarks: manually wound chronograph with 60-minute sweep counter; minutes are displayed in sweep manner where the chronograph seconds would usually be, making them especially legible; a small window at 6 o'clock displays whether the stop function is activated
Estimated value: $1,430 →

1970
Chronograph Datora

Reference number: 2034.3
Case: steel case, tripartite, push-down case back, rotating bezel, Ø 41 mm
Movement: manual winding
Remarks: manually wound chronograph with 45-minute counter; date at 6 o'clock
Estimated value: $1,560 →

1975
Chronograph Sprint

Reference number: 2018
Case: stainless steel, plastic bezel, 40 x 43 mm
Movement: Valjoux Caliber 7733, nickel-plated, manual winding
Remarks: manually wound chronograph with 30-minute counter; polished steel chronograph components; monometallic balance; Incabloc shock protection
Estimated value: $1,300 →

1975
Chronograph Transocean

Reference number: 7102.3
Case: stainless steel, screw down case back, 42 x 47 mm
Movement: Valjoux Caliber 7740, manually wound
Remarks: manually wound chronograph with 30-minute and 12-hour counters
Estimated value: $2,340 →

1975
Chronograph Pupitre

Reference number: 7101.3
Case: steel case, tripartite, screw-down case back, rotating bezel, Ø 42 mm
Movement: Valjoux Caliber 7740, nickel-plated, manual winding
Remarks: manually wound chronograph with 30-minute and 12-hour counters; so-called Bullhead chronograph: by rotating the movement 90 degrees, the buttons for the chronograph functions are positioned at 11 and 1 o'clock, making them thus reminiscent of bull horns
Estimated value: $4,940 ↗

1966
Chronograph Top Time

Reference number: 2006
Case: steel case, tripartite, push-down case back, 36 x 36 mm
Movement: Valjoux Caliber 7730, manual winding
Remarks: simple manually wound chronograph
Estimated value: $1,300 →

CYMA

How is it possible to make life easier for retail watch shops? How could the watchmakers be helped in working more effectively and quickly, ultimately making their lives simpler?

Cyma, one of Switzerland's biggest watch manufacturers with more than 1,000 employees, found an answer. It was one of the first companies to produce completely interchangeable components for its movements in the 1920s. This sensational development was only made possible with precise manufacturing processes: hours of finishing and adjustment during repair were no more, for the company's replacement components fit without problems right into the movements.

Cyma, quickly advancing to one of Switzerland's largest watch companies, already had a tempestuous history behind it at this time. Although it was only officially registered under this name in 1903, its roots reach back to 1862, not to one single company, but a combination of several located in various Swiss regions.

Schwob Frères in La Chaux-de-Fonds is generally pronounced the origin of the later *manufacture*. Schwob Frères was apparently also a financial participant in Sandoz & Cie, which was established in Le Locle in 1871 by Henri Sandoz.

Sandoz moved to Tavannes, north of Biel, in 1891 and specialized in the manufacture of exclusive complication watches, repeaters, and chronographs as Tavannes Watch Co. Using the most modern machines, the company—which distributed its products under the brand names Cyma, Cyma-Tavannes, and even just Tavannes—developed into one of the largest watch manufacturers in Switzerland by 1910.

Even simple Cyma models were very precise watches with excellent rate results and were often sold as officially tested chronometers. Ladies watches, on the other hand, were not available from Cyma for a long time. Their small movements were still too imprecise in the 1920s and, thus, not good enough for Cyma. Only when their small movements ran precisely and reliably did Cyma add a ladies' watch line.

All Cyma models had simple, elegant shapes. The alarm wristwatch Time-O-Vox, which was in the program for a short time, was one of the most popular models. At this time, Cyma used an in-house development for the balance's shock protection called Cymaflex and advertised it extensively.

Alongside beautiful manually wound watches, Cyma also had two automatic caliber families in its program. While the hammer automatics were chiefly used for winding according to the Harwood principle back then, Cyma developed eccentric winding with a spring-loaded rotor axis for its Autorotor caliber.

Very early on, Cyma turned to the groundbreaking microelectronics ravaging the industry for rate precision: in 1973, the company was already producing electronic watches. Thanks to these early tendencies, Cyma survived the Swiss watch industry's quartz crisis in the 1970s. Today, the company works strictly with quartz watches in the private label sector.

Founded:
- 1862 as Schwob Frères & Co.
- 1903 Cyma trademarked in La Chaux-de-Fonds

Headquarters:
La Chaux-de-Fonds, manufacturing in Tavannes

Status:
revived

Specialty:
chronometers, diver's watches, watches for the blind, wristwatch alarms, three-day movements for wristwatches

Major inventions:
- 1890 chronograph patent
- 1960 spring-born rotor axis (Caliber 481)
- around 1948 Cymaflex shock protection

Leading models:
Autorotor, Time-O-Vox

1920
Chronograph

Case: silver, push-down case back, leather strap, Ø 34 mm
Movement: silver-plated, côtes de Genève decoration, column wheel control of chronograph, manual winding
Remarks: rare chronograph with 30-minute counter; enamel dial; signed Tavannes Watch
Estimated value: $2,340 →

1930
Chronomètre

Case: chrome-plated, push-down case back, leather strap, 22 x 37 mm
Movement: silver-plated, côtes de Genève decoration, jewels set in chatons, manual winding
Remarks: chronometer wristwatch
Estimated value: $325 →

1940
Service Watch

Case: stainless steel, screw-down case back, textile strap, Ø 38 mm
Movement: rhodium-plated, côtes de Genève decoration, jewels set in chatons, manual winding
Remarks: wristwatch for the British military forces (signed "W.W.W.")
Estimated value: $520 ↗

1950
Chronomètre

Case: 18-karat yellow gold, push-down case back, leather strap, Ø 34 mm
Movement: nickel-plated, côtes de Genève decoration, jewels set in chatons, manual winding
Remarks: gold chronometer wristwatch
Estimated value: $780 →

1950
Chronomètre

Case: 18-karat yellow gold, push-down case back, leather strap, Ø 33 mm
Movement: rhodium-plated, côtes de Genève decoration, jewels set in chatons, manual winding
Remarks: chronometer wristwatch with original Cyma sales tag
Estimated value: $780 →

1954
Automatic

Case: stainless steel, tripartite, stainless steel screw-down case back, leather strap, Ø 35 mm
Movement: Cyma Caliber 425, gold-plated, polished screws, gold-plated pendulum oscillating weight in jewel bearings, Cymaflex shock protection, 17 jewels (3 set in chatons), automatic
Remarks: very rare Cyma hammer automatic with date display; the movement is outfitted with its own Cymaflex shock protection
Estimated value: $260 →

1953
Time-O-Vox

Case: gold-plated stainless steel, push-down case back, leather strap, Ø 34 mm
Movement: Cyma Caliber 464, gold-plated, gong, manual winding
Remarks: extremely rare alarm wristwatch, only about 3,000 pieces manufactured; subsidiary seconds and sweep alarm hand; watch operated via the crown and two small buttons
Estimated value: $1,300 →

archive photo

1968
Navystar Chronomètre

Case: 18-karat yellow gold, screw-down case back, leather strap, Ø 35 mm
Movement: Caliber 485.1, nickel-plated, côtes de Genève decoration, 27 jewels, automatic
Remarks: gold chronometer wristwatch with Cyma manufacture Caliber 485.1; eccentric rotor
Estimated value: $780 →

DOXA

Georges Ducommun of Le Locle, who had just come of age in 1889, was trying to decide how to best spend his future. He recognized the great need for pocket watches and wanted to catch this train. Therefore, he founded a workshop for watches at the age of twenty-one. His success said it all as the small workshop soon grew into a large factory.

The secret of Ducommun's rapid expansion was the quality of his products: he was known for personally examining every one of the watches to leave his company despite its quick growth. Not only increasing sales numbers showed that he was on the right track; his watches were also awarded several prizes at the exhibitions in Liège in 1905 and Milan in 1906.

In 1908, Ducommun developed an eight-day movement and had it patented right away. It was to be used in the car dashboard timekeepers he had recently added to his program. Since he himself had purchased one of the newfangled iron horses, he was convinced that this invention would be a good sales market for his watch brand as well.

The successful businessman registered the name Doxa in 1910 for a line of his watches; the name was later applied to the whole company. And in the 1920s, he made the first Doxa wristwatches. Shortly thereafter, Ducommun extended his program again: alongside normal, reasonably priced everyday watches, he also manufactured jewelry watches for ladies and gentlemen. He had so much success with his company that he became rich and purchased the chic Château des Monts, a wonderful manor high above the city of Le Locle that today houses the city's watch museum.

In 1936, Ducommun passed away, and his brother-in-law Jacques Nardin took over management of Doxa. He also literally had watchmaking in his blood—for he was the grandson of Ulysse Nardin, the founder of the *manufacture* of the same name in Le Locle.

Nardin continued to expand the company's program. Doxa now manufactured chronographs primarily outfitted with calibers by Valjoux and Venus. There were also unusual watches in the program, among them the 1958 Grafic with its noteworthy shape. The company further introduced a technical specialty in the shape of a simple three-handed watch containing a jump hour display.

The most well-known Doxa model to this day remains one that is equally as uncompromising: the Sub 300 introduced in 1967, a large, heavy diver's watch that is water-resistant to 300 meters and outfitted with an obvious orange dial. The dial color is one that the manufacturer chose on purpose: the company's developmental consultants—members of the U.S. Diver's Company—had found out that under water in the twilight of mid-range diving depths, a watch with an orange dial is most easily read. But the special elements of the Sub 300 don't stop there. Its rotating bezel has two rows of numbers: the dive time can be read from the inner row, while the outer is a functional decompression display, an element that lends this timepiece the predicate Professional. Doxa developed an entire family of diver's watches, among others some with black and silver dials. A chronograph also belonged to the program.

In 1978, Doxa lost its independence and has since belonged to Aubry Frères SA. Recently, the brand has created some wind with successful remake editions of the diver's watches.

Founded:
- 1889 by Georges Ducommun in Le Locle
- 1910 Registry of the trademark Doxa

Headquarters:
Le Locle

Status:
active, after several takeovers

Specialty:
diving watches, designer watches, automobile watches

Major inventions:
1967 orange dials for diving watches for better legibility

Leading models:
Jumping seconds, Sub 300 T, Sub 200 T Graph

1940
Antimagnétique
Case: nickel-plated, push-down case back, leather strap, Ø 38 mm
Movement: nickel-plated, manual winding
Remarks: wristwatch with movable strap lugs
Estimated value: $390 →

1940
Chronograph Antimagnetic
Case: stainless steel, push-down case back, leather strap, Ø 37 mm
Movement: nickel-plated, column-wheel control of chronograph, manual winding
Remarks: chronograph with 30-minute counter
Estimated value: $1,170 →

1940
Chronograph
Case: stainless steel, push-down case back, leather strap, Ø 35 mm
Movement: nickel-plated, manual winding
Remarks: chronograph with off-center display of the time at 12 o'clock
Estimated value: $390 →

1940
Pilot's Chronograph
Case: stainless steel, screw-down case back, leather strap Ø 38 mm
Movement: rhodium-plated, côtes de Genève decoration, column-wheel control of chronograph, manual winding
Remarks: rare pilot's watch for the German Air Force
Estimated value: $1,820 →

1948
Men's Automatic Watch
Case: 18-karat yellow gold, push-down case back, leather strap, 31 x 37 mm
Movement: Caliber AS 1361N, nickel-plated, 17 jewels, automatic
Remarks: automatic men's watch
Estimated value: $320 →

1949
Chronograph with Complete Calendar
Case: stainless steel, push-down case back, leather strap, Ø 35 mm
Movement: Valjoux Caliber 72C, nickel plated, column-wheel control of chronograph, manual winding
Remarks: chronograph with 30-minute and 12-hour counters; complete calendar
Estimated value: $2,600 →

1950
Retrograde Hours Antimagnetic
Case: 14-karat yellow gold, screw-down case back, leather strap, Ø 35 mm
Movement: Caliber 118, nickel-plated, jumping seconds, retrograde hours, manual winding
Remarks: rare wristwatch with retrograde hours (day/night display) and jumping seconds
Estimated value: $1,430 →

1965
Jumping Seconds
Case: stainless steel, push-down case back, leather strap, Ø 35 mm
Movement: nickel-plated, jumping seconds, manual winding
Remarks: extremely rare wristwatch with added mechanism for jumping seconds display; this watch was offered with original Doxa sales tag
Estimated value: $1,170 →

EBERHARD & CO.

In 1887, Georges-Emile Eberhard founded the company Eberhard & Rosselet in La Chaux-de-Fonds. The early watches were sold under the brand names Saphir and later Audax. Eberhard's watches became popular, likely because he often traveled to neighboring countries in order to profit from all the market opportunities there as well. Thanks to his dedication and hard work, he expanded the company in 1906. His specialty was precision watches and chronographs, and soon he was working on his own wristwatch models.

Early on, Eberhard successfully made timekeepers for automobile competitions. In Italy, one of the hot spots for such interests, Eberhard was the official timekeeper for almost all motorized sporting events until well into the 1930s.

From 1919, Eberhard's two sons Georges and Maurice took over the company's management. They recognized that the future of watches was on the wrist. And this was especially true of the world of sports, in which precise timekeeping was a must, and in which the main figures generally needed to keep their hands free. The watchmakers began developing a chronograph for the wrist, and after just one year of research, they introduced the new model: a monopusher chronograph that was to become the brand's flagship model.

Ten years later, Eberhard & Co. became the official watchmaker of the Royal Italian Navy, not in the least because of its contributions to Italian motor sports.

The company continued to consistently develop its sports watches. From 1935, Eberhard & Co. could for the first time boast two-button chronographs that made addition timekeeping possible. From 1938, the company's collection was enriched by a model featuring a 12-hour counter.

In 1939, Eberhard & Co. offered a first split-seconds chronograph as its program's top model. Now the company had practically every variation of sports timekeeping covered.

In the 1940s, a new caliber came out: chronograph Caliber 1600, which was based on a Valjoux *ébauche.* It was essentially a column wheel-controlled monopusher chronograph, but it had a special element: a slide located at 4 o'clock allowed the wearer to interrupt the timing as often as desired and start it up again. Caliber 1600 was therefore very well suited for every kind of addition timekeeping.

In 1942, Maurice Eberhard became the sole owner of the company. In 1958, the company expanded its sports watch line with a professional diver's watch, the Scafodat. In 1962, Eberhard's daughter died in a car accident, leaving him a broken man. He left the business world, selling off the successful venture. Today, Eberhard & Co. is owned by Italians, though it continues to do its manufacturing in Switzerland. It moved its headquarters to Lugano and once again started making headway on the market with models like the Tazio Nuvolari in 1992, the Chrono-4 line (with the four subsidiary dials in a row) in 2001, and a new edition of the Scafodat diver's watch in 2006.

Founded:
1887 by Georges-Emile Eberhard in La Chaux-de-Fonds

Headquarters:
Biel/Bienne (since 1980)

Status:
active without interruption

Specialty:
chronographs, diving watches

Major inventions:
- 1939 in-house development of add function (Caliber 1600)
- 1997 eight-day caliber with a second spring barrel

Leading models:
Extra-fort, Hydro Sub

1925
Chronograph
Case: 18-karat yellow gold, leather strap, Ø 39 mm
Movement: nickel-plated, column-wheel control of chronograph, manual winding
Remarks: chronograph with 30-minute counter; enamel dial; hinged hunter case back
Estimated value: $3,900 →

1945
Split-Seconds Chronograph Extra-Fort
Case: 18-karat yellow gold, push-down case back, leather strap, Ø 40 mm
Movement: rhodium-plated, double column wheel control of chronograph, manual winding
Remarks: chronograph with 30-minute and 12-hour counters; split-seconds mechanism for stopping intermediate times
Estimated value: $10,400 →

1945
Chronograph
Case: 18-karat yellow gold, push-down case back, leather strap, Ø 39 mm
Movement: nickel-plated, column-wheel control of chronograph, manual winding
Remarks: chronograph with 30-minute counter
Estimated value: $3,640 →

1945
Chronograph Extra-Fort
Case: stainless steel, 18-karat gold bezel, push-down case back, leather strap, Ø 40 mm
Movement: Valjoux Caliber 65, nickel-plated, column-wheel control of chronograph, manual winding
Remarks: nearly new chronograph with 30-minute counter
Estimated value: $3,250 →

1948
Chronograph
Case: stainless steel, push-down case back, leather strap, Ø 40 mm
Movement: rhodium-plated, column-wheel control of chronograph, manual winding
Remarks: large chronograph with 30-minute counter
Estimated value: $4,950 →

1950
Chronograph with Complete Calendar
Case: 18-karat yellow gold, push-down case back, leather strap, Ø 35 mm
Movement: rhodium-plated, column wheel control of chronograph, manual winding
Remarks: chronograph with 30-minute and 12-hour counters; complete calendar
Estimated value: $3,250 →

1955
Chronograph Extra-Fort
Case: 18-karat yellow gold, push-down case back, leather strap, Ø 39 mm
Movement: nickel-plated, column-wheel control of chronograph, manual winding
Remarks: chronograph with 30-minute and 12-hour counters
Estimated value: $3,900 →

1956
Chronograph
Case: stainless steel, push-down case back, leather strap
Movement: Caliber 1600, nickel-plated, column-wheel control of chronograph, manual winding
Remarks: chronograph with 30-minute and 12-hour counters; chronograph functions as one-button chronograph; second button at 4 o'clock for additional measurements; pressing this button stops the chronograph, pressing again allows measurement to continue
Estimated value: $4,420 →

archive photo

ENICAR

British race car driver Stirling Moss wore an Enicar. So did his colleague Jim Clark. Diving pioneer Hans Hass counted on the reliability of his Enicar watch, a coup that the company communicated in its advertising of the day.

While people might remember these facts, hardly anyone remembers that the company was founded on a little veranda on Rue du Crêt in La Chaux-de-Fonds, or that the name Enicar is the name of its founder spelled backwards: Racine.

Together with his wife Emma, Ariste Racine founded the company on October 1, 1913. The name was trademarked in 1914. Their first customer was a Japanese man by the name of Tezohdo. Just one year later, the Racines began doing business in Russia. Their most successful models were 11½-line and 13-line calibers outfitted with an unusual additional function: a little compass at 12 o'clock. The brand later made a specialty of watches bearing the portrait of the political leader of the country in which they were going to be sold.

Just two years after its establishment, the company founder experienced some difficulty hiring a qualified workforce in La Chaux-de-Fonds. Thus, he founded a subsidiary in his mother's workshops in Lengnau in 1916 with six employees and called it Longeau Watch.

A new chapter in the history of Enicar began in 1918 when Ariste's brother, Oskar, joined the brand. He was placed in charge of building up the caliber production. Enicar grew quickly; the Racines increased the size of their company in Lengnau and decided to give up the venue in La Chaux-de-Fonds.

The Enicar brand only surfaced in Lengnau in 1932 when Longeau Watch became a joint-stock company, though all of the shares remained in the hands of the family—even in 1940 when Ariste Jr., the founder's son, took over the management of the company that was as old as he was.

It was the above all extremely robust and reliable watches that made the name Enicar famous; they kept on ticking whether under water or high in the Himalayas. An exceptional example of their endurance was the fact that Enicar had outfitted the Swiss Himalaya expedition of 1956 with a new automatic model. The mountain climbers were enthusiastic and loved to compare the rugged reliability of their watches with their sherpas, the native porters who accompanied them into the mountains. Thus, this timekeeper became the original member of the Sherpa model family, whose very solid, sporty timepieces were remarkable in their extreme robustness. It is no accident that the Sherpa models became those most symbolic of the Enicar brand.

The watches were based on Enicar's own caliber series 1120, 1140, and 160. The first caliber family, 1120, was offered in a version called Supertest. These watches were chronometers—models that the company itself had regulated in three positions before selling. In the thirty-jewel automatic version (1125), these were sometimes even offered with a jewel bearing called the Rubyrotor.

Thanks to a solid model policy, it looked as if Enicar would survive the 1970s quartz crisis without damage. But the breakdown of the Asian market hit the company hard and led to Enicar SA's insolvency. It had to declare bankruptcy on July 15, 1987.

Enicar's calibers experienced a somewhat more fortunate destiny: today they continue to tick as the base of many Chronoswiss automatic calibers.

Founded:
1913 by Ariste and Emma Racine-Blatt

Headquarters:
Lengnau

Status:
dissolved, movements used by Chronoswiss

Specialty:
diving watches, sports watches, chronographs, dashboard chronometers, table alarms

Major inventions:
- 1958 oil-embedded bezel on the Sherpa Dive
- 1961 ruby ball-bearings for the rotor
- 1985 Seapearl case with bayonet lock

Leading models:
Sherpa Dive/Divette, Sherpa Guide, Sherpa Graph, Jet Graph

1964
Sherpa Guide Worldtime Rubyrotor

Case: stainless steel, bayonet case back, leather strap, Ø 43 mm
Movement: Caliber AR 1126, gold-plated, jewel bearings, 30 jewels, automatic
Remarks: Caliber 1126 with jewel bearing rotor; an independent, second time can be set via the reference city ring
Estimated value: $1,440 →

archive photo

1969
Jet Graph

Case: stainless steel, screw-down case back, rotating bezel, leather strap, Ø 41 mm
Movement: Valjoux Caliber 724, gold-plated, column wheel control of chronograph, manual winding
Remarks: chronograph with 30-minute and 12-hour counters; second time zone set via the bezel; in addition, an additional time zone can be set via the rotating small red triangle reference marker
Estimated value: $2,100 →

1969
Sherpa Jet Automatic

Case: stainless steel, screw-down case back, leather strap, Ø 36 mm
Movement: nickel-plated, automatic
Remarks: sporty wristwatch with two time zones (GMT) and two-tone 24-hour ring
Estimated value: $1,170 →

1970
Super Dive Sherpa

Case: stainless steel, screw-down case back, leather strap, Ø 40 mm
Movement: Caliber AR 1145, gold-plated, automatic
Remarks: Sherpa diver's watch with rotating inner bezel (flange)
Estimated value: $1,170 →

1975
Sherpa Jet Automatic

Case: stainless steel, screw-down case back, leather strap, 37 x 41 mm
Movement: Caliber AR 166 (respective 1146), gold-plated, automatic
Remarks: Sherpa with second time zone and colorful 24-hour ring
Estimated value: $1,170 →

1975
Graphomatic

Case: stainless steel, tripartite, stainless steel screw-down case back, leather strap, 44 x 39 mm
Movement: Lemania Caliber 1340, rhodium-plated, fine matte steel components, eccentric fine adjustment, 17 jewels, automatic with winding rotor
Remarks: automatic chronograph with sweep 60-minute counter and small 12-hour counter at 6 o'clock; under-crystal ring with tachymeter scale
Estimated value: $910 →

1975
Alarm

Case: stainless steel, screw-down case back with membranous gasket, leather strap
Movement: Caliber AS 1931, red gold-plated, twin spring barrels, manual winding
Remarks: wristwatch alarm with date and double sounding board case back according to the Vulcain Cricket principle
Estimated value: $650 →

1971
Memostar

Case: stainless steel, screw-down case back, leather strap, 39 x 44 mm
Movement: Lemania Caliber LWO 2980, nickel-plated, gong, automatic
Remarks: rare automatic alarm wristwatch with Lemania caliber outfitted with one single spring barrel
Estimated value: $1,820 →

ETERNA

He was a man of action, and his adventurous journeys brought him global fame—of course, we are talking about Norwegian anthropologist Thor Heyerdahl. In 1947, he left Peru on his balsa wood raft called *KonTiki* in order to test whether people could have immigrated to the Polynesian islands in this manner. Three and a half months and almost 5,000 miles later he reached the Tuamotu Islands. His experiment had worked.

An indispensable piece of equipment for him on this trip was a watch manufactured in small series developed especially for the researcher, an indestructible timepiece whose name was to become a hallmark for its manufacturer: the KonTiki by Eterna.

By this time, the company had already had a turbulent history. It was first founded as Watch Manufacture Urs Schild by a schoolteacher of the same name and Josef Girard, a doctor, in Grenchen, one of the poorest regions of Switzerland. Girard had previously established a watchmaker school and had also shipwrecked an *ébauche* factory he had started up prior to that.

The duo built the new watch *manufacture* inaugurated in 1856 directly on the banks of the city's stream. Thus, they were able to harness the power of its water for their company's use. Success was not long in coming: in 1870 they already had to expand.

Schild introduced modern machines to the factory. His son Max, who had taken over its management in 1888, was in favor of rationalization. He had studied industrial production in the United States at the Illinois Watch Company and wanted to apply those experiences to his own operations. However, the Swiss were afraid of losing their jobs and called a strike. Max was forced out of the company, and his brother Theodor took the rudder. The machine engineer modernized the factory and introduced a new company name in 1906: Eterna-Werke, Gebrüder Schild & Co.

Early on, Theodor saw that the future belonged to the wristwatch and reacted correspondingly. In 1914, he introduced a sensation: the first ever serially manufactured alarm wristwatches. By 1930, Eterna was making the smallest wristwatch produced in series with a baton movement.

Two years later Theodor Schild divided his company in two in response to pressure from the industrial association: Eterna AG for assembling watches and ETA SA specializing in the manufacture of *ébauches.* What continued to connect the two divisions was the fact that all newly developed movements were first put into Eterna watches before they could be sold to other companies.

The movement developed for Heyerdahl and the KonTiki watch in 1947 already possessed a new type of automatic winding whose rotor rolled on five ball bearings, each of which had a diameter of only 0.65 mm. These five balls were to become the brand's logo, and the system became the standard for the entire watch industry.

In the 1970s, Eterna competed with Japanese companies for the slimmest quartz movement—and won with a height of under one millimeter and a weight of just over one gram.

But the quartz crisis was also having its corrosive effect on the company, and only a consolidation within the Swiss watch industry could save Eterna, which became part of ASUAG. Omega and Tissot, part of a group called SSIH, which later became the Swatch Group, took over ETA SA. Eterna AG was sold to the PCW Group in 1984, and then, in 1995, to the F.A. Porsche family. In 2011, a bigger fish came along, the Chinese holding China Haidian, to take possession of Porsche and Eterna, and a year later of Corum.

Founded:
in 1856 by Urs Schild and Dr. Josef Girard in Grenchen

Headquarters:
Grenchen

Status:
active without interruption

Specialty:
sports and diving watches, wrist alarms

Major inventions:
- 1914 wristwatch alarms
- 1948 ball bearing-mounted rotor oscillating mass, springless pawls
- 1980 Marketing of the Linea IV, technological predecessor of the Swatch

Leading model:
KonTiki

1945
Chronograph Doctor's Watch

Case: stainless steel, push-down case back, leather strap, Ø 33 mm
Movement: nickel-plated, column wheel control of chronograph, manual winding
Remarks: nearly unworn chronograph with 30-minute counter and pulsometer scale
Estimated value: $780 →

1950
Chronograph Doctor's Watch

Case: 18-karat yellow gold, screw-down case back, leather strap, Ø 38 mm
Movement: Caliber E 704, nickel-plated, column wheel control of chronograph, manual winding
Remarks: heavy gold chronograph with 30-minute and 12-hour counters and pulsometer scale
Estimated value: $2,080 →

1945
Chronograph with Complete Calendar

Case: 18-karat yellow gold, push-down case back, leather strap, Ø 36 mm
Movement: nickel-plated, column wheel control of chronograph, manual winding
Remarks: chronograph with 30-minute and 12-hour counters; complete calendar
Estimated value: $2,600 →

1945
Chronograph with Complete Calendar

Case: 18-karat yellow gold, push-down case back, leather strap, Ø 36 mm
Movement: nickel-plated, column wheel control of chronograph, manual winding
Remarks: chronograph with 30-minute and 12-hour counters; complete calendar
Estimated value: $520 →

1950
Chronograph

Case: stainless steel, screw-down case back, leather strap, Ø 38 mm
Movement: Caliber 703, rhodium-plated, column wheel control of chronograph, manual winding
Remarks: chronograph with 30-minute counter; telemeter scale and snail-shaped tachymeter scale
Estimated value: $910 →

1965
Eterna-Matic Chronometer Centenaire

Case: 18-karat yellow gold, leather strap, Ø 34 mm
Movement: Caliber 1429U, nickel-plated, automatic
Remarks: anniversary chronometer from 1965
Estimated value: $1,560 →

1968
KonTiki 20

Case: stainless steel case, screw-down case back, stainless steel link bracelet, Ø 36.5 mm
Movement: Eterna Caliber 1489K, rhodium-plated, decorated, 21 jewels, automatic
Remarks: a typical sporty men's watch, the KonTiki was extremely popular in the 1960s
Estimated value: $390 →

archive photo

1965
Eterna-Matic Chronometer

Case: stainless steel, screw-down case back, Ø 34 mm
Movement: Caliber 1422, nickel-plated, decorated, 21 jewels, automatic
Remarks: chronometer wristwatch
Estimated value: $650 →

EXCELSIOR PARK

Jules-Frédéric Jeanneret founded a watchmaking business in 1866 in Saint-Imier and focused on the making of stopwatches. Twenty-two years later, in 1888, Jeanneret's sons took over. One of them, Albert Jeanneret, registered a patent for a chronograph in 1891. Two years later, he transferred the direction of the company to his own sons, Henri and Constant. These brothers, however, had no consensus about how the company should be run. In 1901, they therefore decided to go their separate ways: Constant left and a short time later bought Léonidas, which also specialized in manufacturing chronographs.
Henri Jeanneret continued to manage the company, which was now called Jeanneret-Brehm & Cie. Under his leadership, it grew rapidly. He began using the name Excelsior for the chronographs, then, at some point in time, Excelsior Park, which later became the name of the company as a whole. The name Excelpark is also seen occasionally.
In order to fulfill the high demands that Jeanneret had set, Excelsior Park chiefly developed and manufactured its own calibers for its chronographs. The shaped movement Caliber 42, like all other movements by Excelsior Park, was outfitted with an unusual module: the crown wheel was positioned directly above the balance, making it different from most conventional chronograph calibers. Excelsior Park did not exclusively use its own calibers, however. The company's watchmakers put Venus Caliber 179 containing a split-seconds mechanism, Venus 140 featuring an off-center time display, the modern Valjoux Caliber 7740, and various Landeron calibers in their cases. Although the company specialized in sports chronographs, Excelsior Park also naturally included three-handed watches in its program.
Like many other companies, this chronograph specialist also felt the effects of the quartz crisis at the end of the 1970s. In 1984, Excelsior Park was liquidated. Flume, well-known as a tool supplier to watchmaker workshops based in Essen, Germany, bought the rights to the brand. At first, Flume tried to resuscitate the line and manufactured several small series of chronographs under the Excelsior Park name. The attempt failed, however. Today, the rights to the trademark are back in Switzerland.

Founded:
in 1866 by Jules-Frédéric Jeanneret

Headquarters:
St.-Imier

Status:
dissolved

Specialty:
chronographs, rattrapante chronographs

Major invention:
1891 patent for a chronograph

Leading models:
Excel-O-Graph, Monte Carlo

1948

Chronograph

Case: stainless steel, screw-down case back, leather strap, Ø 37 mm
Movement: nickel-plated, côtes de Genève decoration, column wheel control of chronograph, manual winding
Remarks: chronograph with 45-minute counter
Estimated value: $1,300 ➔

1948

Chronograph

Case: 18-karat rose gold, push-down case back, leather strap, Ø 35 mm
Movement: nickel-plated, côtes de Genève decoration, column wheel control of chronograph, manual winding
Remarks: chronograph with 45-minute counter
Estimated value: $2,080 ➔

ca. 1950

Chronograph Pulsometer

Case: stainless steel, screw-down case back, leather strap, Ø 37 mm
Movement: nickel-plated, côtes de Genève decoration, manual winding
Remarks: chronograph with 45-minute counter and pulsometer scale
Estimated value: $1,560 ➔

1975

Chronograph Monte Carlo

Case: stainless steel, screw-down case back, rotating bezel, leather strap, Ø 42 mm
Movement: Valjoux Caliber 7740, rhodium-plated, manual winding
Remarks: chronograph with 30-minute and 12-hour counters
Estimated value: $1,560 ➔

1970

Excel-O-Graph

Case: stainless steel, push-down case back, rotating bezel, leather strap, Ø 42 mm
Movement: rhodium-plated, côtes de Genève decoration, column wheel control of chronograph, manual winding
Remarks: nearly unworn chronograph with 30-minute and 12-hour counters; rotating bezel can be used as slide rule
Estimated value: $1,950 ➔

ca. 1965

Chronograph

Case: stainless steel, screw-down case back, rotating bezel, leather strap, Ø 39 mm
Movement: Caliber 40-68, rhodium-plated, côtes de Genève decoration, column wheel control of chronograph, manual winding
Remarks: chronograph with 30-minute and 12-hour counters
Estimated value: $2,340 ➔

1980

Chronograph Limited Edition

Case: 18-karat yellow gold, push-down case back, leather strap, Ø 37 mm
Movement: rhodium-plated, côtes de Genève decoration, manual winding
Remarks: unworn chronograph from a limited edition, No. 7/27; 30-minute counter and off-center display of the time
Estimated value: $4,550 ➔

ca. 1990

Chronograph Série Limitée

Case: 18-karat yellow gold, tripartite, exhibition push-down case back, leather strap, Ø 39 mm
Movement: Venus Caliber 185, rhodium-plated, fausses côtes, finely polished, beveled steel components, twin column wheel chronograph control and split-seconds function, manual
Remarks: nearly new limited edition (1/15) with 30-minute counter; split-seconds control via crown-integrated button
Estimated value: $9,100 ➔

FORTIS

In 1912, watchmaker Walter Vogt decided to found a collective for producing watches called Vogt und Rüefli with his partner Alfred Rüefli. Trained as a watchmaker at Eterna, Vogt had the best prerequisites to manufacture reliable timepieces, and he had a genuine desire to aim for top quality. Since, at this time, movement manufacturers only delivered *ébauches* to be assembled by each watch manufacturer—who also added the escapement and even sometimes jewels and the mainspring—it was up to every company how lavishly the movement was to be finished. For Vogt, the priority was clearly the precision of the watches.

By the following year, the brand Fortis had been registered. In 1914, Vogt and Rüefli separated, and the company was renamed Vogt & Co. With the later change to a joint-stock company, it finally took on its current name: Fortis Uhren AG.

Walter Vogt made his most important decision when he elected to manufacture the new automatic wristwatch developed by Englishman John Harwood in the 1920s. Thanks to his good contacts with movement factory A. Schild, also located in Grenchen, both companies began to perfect Harwood's design, optimizing it for serial production. Fortis produced its first chronographs in 1937. The first water-resistant chronograph was named Wandfluh after Grenchen's own mountain. A specialty of the brand was to press the chronograph movements into the case from the front.

A new era began in 1948 when Rolf Vogt joined his father's company. Thanks to his sharp business acumen and feel for new trends, he managed to establish the Fortis brand in many world markets.

Rolf Vogt wanted to give his brand its own look. Thus, a number of exceptional watches such as Eden Roc and Trueline were created, models that are considered design classics today. In 1956, Fortis also had an alarm in its collection, and the Manager introduced two years later was the first water-resistant alarm for the wrist even delivered as an officially certified chronometer.

In order to get the best rate results possible during serial production, Fortis's technicians developed a new quality control element. BEP, or the Balance Electrotiming Process, allowed a watch to produce very impressive accuracy. A timepiece subjected to this process received a special tag to alert the customer to its high quality. The Performance model opened a new dimension in tying the brand to aviation, a relationship that is still valid today, and especially to military aviation. A collection of pilot's watches and chronographs was continuously extended, characterizing the brand's new image along with a line of Cosmonaut watches made for space. This brand continues to be headquartered in Grenchen, though it is now owned by a German family.

Founded:
1912 by Walter Vogt and Alfred Rüefli in Grenchen

Headquarters:
Grenchen

Status:
active without interruption

Specialty:
wristwatches, diving watches, service watches, chronographs, pilot's watches

Major inventions:
- 1926 watch winder to control winding performance
- 1968 Flipper, the first water-resistant watch in a plastic case
- 1998 chronograph with alarm function

Leading models:
Wandfluh, Stratoliner, Marinemaster, Manager

1937
Wandfluh

Reference number: 189 190
Case: stainless steel, screw-down case back, leather strap
Movement: Venus Caliber 152, screw balance, manual winding
Remarks: chronograph with 30-minute and 12-hour counters; rare case design with integrated movement container that is pressed into the case from behind
Estimated value: $1,560 ➔

archive photo

1940
Chronograph

Reference number: 87910
Case: chrome-plated, push-down case back, leather strap, Ø 35 mm
Movement: Venus Caliber 150, column-wheel control of chronograph, manual winding
Remarks: chronograph with 30-minute counter
Estimated value: $520 ➔

archive photo

1940
Chronograph

Reference number: 39804
Case: 18-karat yellow gold, leather strap
Movement: Valjoux Caliber 22, manual winding
Remarks: chronograph with telemeter and inner tachymeter scales, 30-minute counter
Estimated value: $1,820 ➔

archive photo

1952
Chronograph

Case: stainless steel, case container design, push-down case back, leather strap, Ø 36 mm
Movement: Valjoux Caliber 72, Breguet hairspring, rhodium-plated, 17 jewels, manual winding
Remarks: chronograph with 30-minute and 12-hour counters; rare case design with integrated movement container that is pressed into the case from behind
Estimated value: $1,820 ➔

archive photo

1954
Manager

Case: stainless steel, screw-down case back, leather strap,
Movement: Caliber AS 1475, rhodium-plated, twin spring barrels, 21 jewels, manual winding
Remarks: first water-resistant alarm wristwatch, this model was also available as a chronometer
Estimated value: $560 ➔

archive photo

1972
Marinemaster

Reference number: 8001
Case: stainless steel, screw-down case back, rotating bezel, leather strap
Movement: Valjoux Caliber 72, 17 jewels, manual winding
Remarks: chronograph with minutes and hours; rotating black diver's bezel
Estimated value: $1,300 ➔

archive photo

1970
Marinemaster

Reference number: 337462 8-76
Case: stainless steel, screw-down case back, leather strap
Movement: ETA Caliber 2783, 21 jewels, automatic
Remarks: diver's watch with rotating decompression scale on the dial; date display; rotating ring underneath the crystal; luminous dial
Estimated value: $1,040 ➔

archive photo

ca. 1970
Easy-Math

Reference number: 7242
Case: chrome-plated, rotating inner bezel (flange), leather strap, Ø 37 mm
Movement: FHF Caliber 969, rhodium-plated, 17 jewels, manual winding
Remarks: sporty men's watch with compass function and logarithmic calculation scale on the adjustable numeral ring
Estimated value: $780 ➔

archive photo

GIRARD-PERREGAUX

The history of the Girard-Perregaux brand officially began with a marriage: in 1854 Constant Girard and Marie Perregaux decided to spend the rest of their lives together. Two years later, he renamed his company, which had been known as Girard & Cie until then.
The roots of the company, however, extend even further back, all the way to the year 1791. At that time watchmaker Jean-François Bautte joined the company of his teacher, J. D. Moulinier. The enterprise grew swiftly and well, and it was mainly blue bloods who swore on the timekeepers of Moulinier & Bautte. When Bautte died in 1837, his grandson took over the company and sold it to Felipe Hecht in 1897. It was only in 1905 that his son Juan sold it to the owners of Girard-Perregaux.
Constant Girard dedicated himself to the highest precision for the manufacture of his watches, a plan that his wife Marie, the daughter of a chronometer maker, supported as far as possible. They were also exceptional aesthetes: technical solutions had to be visually pleasing as well. And because Girard had a thing for gold, he even had bridges, wheels, and entire base plates made of the valuable metal. The two could soon harvest the fruits of their labors: they won several chronometer competitions held by the Neuchâtel Observatory, and by 1889 they had received a total of thirteen gold medals and awards.
The company received the last of these for its most famous timepiece: a tourbillon with three parallel gold bridges. Only twenty pieces of this model, the production of which was extremely difficult, were created over the course of twenty-five years. However, its unusual design was rediscovered in the late twentieth century when Girard-Perregaux realized the tourbillon within the framework of a wristwatch.
The *manufacture's* watches were always known as reliable, something that brought the company a lucrative deal with the German navy in 1879. These models, outfitted with a protective grille over the sensitive crystal, were made especially for officers and were worn on the wrist, held by a leather strap. Thus, this model can be said to be one of the oldest wristwatches known to history.
Although Girard-Perregaux was a standing member of the jury at world's fairs since 1905, the brand itself went quiet until it was purchased by Graef & Co.
The Olimpico series was one of Girard-Perregaux's most traditional model lines. Every four years the company presented a new model in honor of the Olympic Games.
The automatic watches sold under the name Gyromatic were more well-known, however. This line received its name from the special elements that allowed the rotor to wind the mainspring in both rotational directions. Small balls called *gyrotrons* moved in three radially arranged conical slits in the minute wheels.
Today, it is less well-known that Girard-Perregaux was one of the pioneers of quartz technology, even defining the standard frequency of 32,768 hertz. In 1970, the company brought the first serially manufactured quartz wristwatch to the market.
Parallel to this, the company continued working on increasing the precision of mechanical movements: thus, the HF, an automatic watch with an ultra-fast frequency, was born. Its balance beat at 36,000 half-oscillations per hour.
Girard-Perregaux is one of the few companies to retain its *manufacture* status. In 1992, it was sold to Turin entrepreneur and former race car driver Dr. Luigi Macaluso, who continued expanding the production depth. After his death in October 2010, the majority shareholder became Kering Group (formerly PPR).

Founded:
1852 by Constant and Numa Girard

Headquarters:
La Chaux-de-Fonds

Status:
active without interruption
(Sowind Group today)

Specialty:
chronometers, tourbillons, jaquemarts, minute repeaters

Major inventions:
- 1879 design of a wristwatch for the German navy
- 1880 chronograph with 30-second foudroyante
- 1889 tourbillon under three bridges
- 1957 Gyromatic
- 1965 high-frequency movement with 36,000 vph
- 1969 quartz standard 32,768 kHz

Leading models:
Tourbillon under Three Gold Bridges
high-frequency chronometer (36,000 vph)

1940

Doctor's Chronograph

Case: 14-karat yellow gold, push-down case back, leather strap, Ø 31 mm
Movement: Caliber 281, nickel-plated, column wheel control of chronograph, manual winding
Remarks: so-called doctor's chronograph with scale for easy measurement of the pulse
Estimated value: $1,820 ➔

1945

Men's Watch

Case: chrome-plated, push-down case back, leather strap, Ø 35 mm
Movement: rhodium-plated, manual winding
Remarks: unworn men's watch
Estimated value: $260 ➔

1945

Men's Watch with Large Date

Case: stainless steel, push-down case back, leather strap, 24 x 37 mm
Movement: nickel-plated, jewels set in chatons, manual winding
Remarks: rare men's watch with large date; date changed by pushing the crown
Estimated value: $650 ➔

1950

Calendar Watch

Case: stainless steel, screw-down case back, leather strap, Ø 33 mm
Movement: Caliber 382, nickel-plated, automatic
Remarks: men's watch with subsidiary seconds and complete calendar (date hand, display windows for weekday and month)
Estimated value: $390 ➔

1960

Chronometer GP

Case: 18-karat yellow gold, screw-down case back, leather strap, Ø 34 mm
Movement: rhodium-plated, regulated in 5 positions, manual winding
Remarks: chronometer with patented fine adjustment; movement regulated in 5 positions
Estimated value: $1,430 ➚

1965

Chronometer HF

Reference number: 8856B
Case: stainless steel, screw-down case back, leather strap, Ø 34 mm
Movement: Caliber 30, nickel-plated, manual winding
Remarks: unworn chronometer with extremely high frequency (36,000 vph)
Estimated value: $1,040 ➚

1970

Men's Watch

Case: stainless steel, push-down case back, leather strap, Ø 37 mm
Movement: rhodium-plated, manual winding
Remarks: unworn elegant men's watch with subsidiary seconds
Estimated value: $260 ➔

1975

Chronograph Olimpico

Case: stainless steel, screw-down case back, rotating bezel, leather strap, Ø 39 mm
Movement: rhodium-plated, decorated, manual winding
Remarks: chronograph with 30-minute and 12-hour counters
Estimated value: $780 ➔

GLASHÜTTE

If you research the roots of the former combine Glashütter Uhrenbetriebe, you will come across names that read like a *Who's Who* of the German watch industry. The conglomerate's direct predecessor was Deutsche Präzisions-Uhren-Fabrik Glashütte GmbH, founded in 1918. Due to economic problems, this company was split in 1925 and became Uhren-Rohwerke-Fabrikation Glashütte AG (Urofa) and Glashütter Uhrenfabrik AG (Ufag). Glashütter Uhrenbetriebe was later created for the most part from these two firms in the German Democratic Republic.

Ufag sold its watches under the brand name Tutima, while Urofa only made movements like the famed Caliber 59 from about 1941, a 13-line chronograph movement with flyback function.

The most well-known brand to be found in the Glashütter Uhrenbetriebe combine was A. Lange & Söhne. Its founder, Ferdinand Adolph Lange, was the man who brought watchmaking to the economically depressed Erzgebirge region, making Glashütte the most important German center of the watchmaking industry at the time. Glashütte also had him to thank for the many small companies that found their home in the region including suppliers, assemblers, and other specialists. Among the most important of those were Adolf Schneider, Moritz Grossmann, Ludwig Strasser, Gustav Rohde, and Julius Assmann with his Union watch factory. The Mühle manufactory, which was well-known for its marine chronometers, was also later integrated into Glashütter Uhrenbetriebe.

After World War II, Glashütte's production workshops were dismantled as reparations to the Soviet Union and shipped eastward. But by 1946, Glashütte was once again manufacturing watches—and these were not to be sneezed at: the calibers were offered in various quality levels, with Q1 being the highest.

The real history of Glashütter Uhrenbetriebe began in 1951 when, in the GDR's Glashütte, brands were put together to form the VEB Glashütter Uhrenbetriebe combine. One of Glashütte's specialties became the rare chronograph Caliber 64. Several manually wound and automatic movements were also manufactured there.

In the 1960s, Glashütter Uhrenbetriebe supplied many of the so-called catalogue watches for West German mail-order companies such as Quelle. These watches were outfitted with Spezichron and Spezimatic calibers driven by automatic winding or quartz movements. They were also 100-percent Glashütte design and manufacture and an important source of Western currency in the GDR.

With the end of the German Democratic Republic in 1989, Glashütter Uhrenbetriebe also had to be restructured. The basis of the East's upswing was privatization, and the now-privatized company founded a new brand called Glashütte Original, which based its production on the newly developed automatic Caliber 10-30 that came about shortly before the fall of the Berlin Wall.

The Glashütte watchmakers left in the current incarnation of the company now continually conceive and manufacture new complications showcasing the talent of this new-old *manufacture*. In 2000, Glashütter Uhrenbetriebe was sold to Switzerland's Swatch Group.

Founded:
1951 as VEB Glashütter Uhrenetriebe

Headquarters:
Glashütte in Saxony

Status:
active without interruption (today Glashütter Uhrenbetrieb in the Swatch Group)

Leading models:
Spezichron, Spezimatic

1937
Tutima Men's Watch

Case: gold-plated, stainless steel push-down case back, leather strap
Movement: Caliber 58, 16 jewels, manual winding
Remarks: Tutima men's watch with shaped movement developed by Urofa in 1935 and available in different sizes
Estimated value: $520 ➔

archive photo

1942
Tutima Chronograph

Case: chrome-plated, push-down case back, leather strap
Movement: Caliber 59, nickel-plated, column wheel control of chronograph, 17 jewels, manual winding
Remarks: so-called German army chronograph outfitted with Urofa Caliber 59 (official supply watch)
Estimated value: $5,200 ➔

archive photo

1945
Urofa Men's Watch

Case: chrome-plated, stainless steel push-down case back, leather strap, Ø 35 mm
Movement: Caliber 611, nickel-plated, manual winding
Remarks: rare Urofa men's watch
Estimated value: $325 ➔

1950
Men's Watch Q1

Case: 14-karat yellow gold, leather strap, Ø 33 mm
Movement: Caliber 28, gold-plated, frosted finish, manual winding
Remarks: rare yellow gold Quality 1 men's watch
Estimated value: $2,600 ➔

1960
Chronograph

Case: gold-plated, stainless steel screw-down case back, leather strap, Ø 36 mm
Movement: Caliber 64, nickel-plated, column wheel control of chronograph, manual winding
Remarks: extremely rare chronograph with 30-minute counter; this design is based on the so-called German army chronograph, Caliber 59, and has a much smaller diameter
Estimated value: $1,950 ➔

ca. 1955
Chronometer Q1

Case: gold-plated, stainless steel push-down case back, leather strap, Ø 36 mm
Movement: Caliber 70.3, nickel-plated, manual winding
Remarks: rare wristwatch chronometer with Quality 1 movement
Estimated value: $1,040 ➔

1955
Spezimatic

Case: yellow gold, push-down case back, leather strap, Ø 34 mm
Movement: Caliber 662, rhodium-plated, automatic winding
Remarks: rare gold automatic watch
Estimated value: $1,040 ➔

1955
Spezimatic

Case: 14-karat yellow gold, push-down case back, leather strap, Ø 34 mm
Movement: Caliber 661, rhodium-plated, 26 jewels, automatic winding
Remarks: rare gold automatic watch with date
Estimated value: $1,040 ➔

HANHART

Whether it was during gym class or a sports competition, almost every European kid has had to rely on what a Hanhart stopwatch has said about his or her athletic skills at some point. This company hailing from Gütenbach in the Black Forest was long one of the most reputable and biggest manufacturers of mechanical and electronic stopwatches. Its roots extend almost 125 years back into the past—and into Switzerland.

There—or, more precisely, in Diessenhofen—is where Johann Adolf Hanhart opened his first watch and jewelry shop in 1882. Twenty years later, he moved to Schwenningen in the Black Forest. Upon joining the company in 1920, his son Wilhelm Julius (Willy) recognized an opportunity and invested in the design of a reasonably priced stopwatch. Four years later, he laid the cornerstone for an exceptional career in the world of sports and began his tenure as the world's best-known and most important stopwatch producer.

From 1932 on, his company dedicated itself solely to the manufacture of watches. The first autonomous pocket and wristwatch calibers were quickly developed, including stopwatch Caliber 57 with its ultra-fast frequencies of 180,000 and 360,000 vph. This watch made precision measurements of up to one-hundredth of a second possible.

The most famous model of the Black Forest company was created in 1939: a pilot's chronograph with flyback function. The resetting of the chronograph "on the fly" with immediate restarting made it possible for pilots to keep precise time while navigating. During World War II, the Hanhart chronograph was chosen as official equipment for German pilots. Hanhart paid the price for this honor after the war: in 1945 the company was dismantled. Willy Hanhart himself was threatened with incarceration, despite his status as a Swiss citizen. He fled to his country of birth in 1947, where he remained until 1949.

During this upheaval, his wife reorganized the company in Gütenbach. Hanhart once again had an extensive program containing pocket and wristwatches, even continuing to make the famous pilot's watch until 1962.

In 1951, the company introduced another of its own developments: the Sans-Souci alarm wristwatch powered by Caliber 301. At the same time, the brand began to offer a civilian version of the famous chronograph. In the 1950s, there was a large selection of different wristwatches in the collection.

Hanhart survived the quartz crisis more or less thanks to its stopwatches, but ran into financial trouble at the beginning of the 1990s. In 1992, a group of Munich-based investors acquired the venerable company and put it back on a sound road.

Today, alongside stopwatches, replicas of the old pilot's chronographs are the main items associated with the name Hanhart. Both the one-button version and the variation with two buttons asymmetrically arranged with regard to the crown are once again being produced, though these days they house modern movements. In 2010, the Swiss Gaydoul Group took over the company.

Founded:
1882 by Johann A. Hanhart in Diessenhofen, Switzerland

Headquarters:
Gütenbach (Germany) and Diessenhofen (Switzerland)

Status:
active without interruption

Specialty:
stopwatches, chronographs, pilot's watches, wristwatches, wristwatch alarms

Leading models:
pilot's chronograph, Sans-Souci

1935
German Air Force Chronograph

Case: stainless steel, screw-down case back, leather strap, Ø 41 mm
Movement: nickel-plated, column-wheel control of chronograph, manual winding
Remarks: rare German pilot's chronograph with an unusual dial
Estimated value: $3,250 ➔

1940
German Air Force Chronograph

Case: nickel-plated, screw-down case back, rotating bezel, leather strap, Ø 39 mm
Movement: nickel-plated, column wheel control of chronograph, manual winding
Remarks: one-button chronograph of the early German air force period
Estimated value: $2,600 ➔

1940
German Navy Chronograph

Case: nickel-plated, screw-down case back, rotating bezel, leather strap, Ø 39 mm
Movement: nickel-plated, column wheel control of chronograph, manual winding
Remarks: rare one-button chronograph of the German navy with stamped eagle and navy number (KM185)
Estimated value: $4,550 ➔

1942
German Air Force Chronograph

Case: nickel-plated, screw-down case back, rotating bezel, leather strap, Ø 39 mm
Movement: nickel-plated, column wheel control of chronograph, manual winding
Remarks: rare one-button pilot chronograph of the German air force with rotating bezel
Estimated value: $2,340 ➔

1942
German Air Force Chronograph

Case: stainless steel, screw-down case back, rotating bezel, leather strap, Ø 38 mm
Movement: nickel-plated, column wheel control of chronograph, manual winding
Remarks: stainless steel pilot's chronograph with 30-minute counter
Estimated value: $2,340 ➔

1944
German Air Force Chronograph

Case: chrome-plated, screw-down case back, leather strap, Ø 41 mm
Movement: nickel-plated, column wheel control of chronograph, manual winding
Remarks: one-button pilot's chronograph of the early German air force period with 30-minute counter
Estimated value: $2,340 ➔

1952
Sans-Souci

Case: chrome-plated, stainless steel push-down case back, rotating bezel, leather strap, Ø 36 mm
Movement: Hanhart Caliber 301, nickel-plated, 17 jewels, manual winding
Remarks: wristwatch alarm with one spring barrel; this watch was owned by movement designer Malonek who developed Caliber 301 into Caliber 302 for Hanhart; Malonek wore this watch for testing as well as on private trips; has a deployable resonance bridge
Estimated value: $520 ➔

1952
Sans-Souci

Case: gold-plated, stainless steel push-down case back, rotating bezel, link bracelet, Ø 36 mm
Movement: Hanhart Caliber 301, nickel-plated, manual winding
Remarks: wristwatch alarm with one spring barrel; this watch comes from the second series and has a slide at 9 o'clock to control the alarm function; has a deployable resonance bridge
Estimated value: $390 ➔

HEUER

Also known as Léonidas, Heuer-Léonidas, and TAG Heuer

In 1860, twenty-year-old Edouard Heuer founded a watch workshop in Saint-Imier. His company grew rapidly, so that only four years later he needed to look for larger facilities, which he found in Biel. In 1880, Heuer produced his first serial chronographs—and by the time he passed away in 1892, his watchmakers were already considered specialists in chronographs of every kind.

His sons Jules and Charles managed the company starting in 1902. They presented a so-called doctor's chronograph in 1910 with which doctors could calculate their patients' pulse rates to within just a few seconds. The brand's first wristwatches left the factory after 1913—at first, however, these were only ladies' models. In 1916 Heuer patented a sensation: the Microsplit, a stopwatch that could measure the time precisely to the hundredth of a second.

The chronograph business developed into the main source of income for Heuer in the ensuing years. The company bought movements, encased them, and gave them additional functions such as slide rules and tide indicators. As of 1933, the company started making dashboard clocks for race and rally cars.

In the mid-1960s, Heuer was one of the main forces behind the development of the first automatic chronograph caliber—legendary Caliber 11—created in conjunction with Breitling, Hamilton-Büren, and Dubois Dépraz. Heuer maintained a large selection of different watch types. An army pilot's chronograph, the Monaco, and the Autavia are of course some of the better-known models. One of Heuer's greatest competitors was Léonidas SA in Saint-Imier, a company that also made chronographs. Julien Bourquin founded this company in 1841, choosing the legendary Spartan king Leonidas, who in antiquity fought against the Persians and then died on the battlefield, as its patron.

At the beginning Léonidas produced simple pocket watches, though Bourquin soon specialized in the manufacture of stopwatches and chronographs. His customers included various foreign ministries, which he supplied with pilot's chronographs and service watches. Alongside that, Léonidas also produced civilian chronographs with calendar functions, three-handed watches with moon phase displays, and alarm wristwatches. Additionally, the company manufactured board clocks for airplanes and automobiles. The most famous of Léonidas's watches, however, remains the pilot's flyback chronograph containing Valjoux Caliber 222. Only 250 of these were made.

On January 1, 1964, Heuer forced the irksome competitor into a merger and at first produced different models under both names. The Léonidas name quickly disappeared from the dials, however.

In order to guarantee continued existence of the Heuer brand during the quartz crisis, Jack W. Heuer saw no other solution than to sell his company to Lemania and a few other stockholders, including the Piaget family. In 1985, Heuer-Léonidas was purchased by the high-tech concern Techniques d'Avant-Garde (TAG), and its name was changed to TAG Heuer.

In 1999, luxury goods concern LVMH (Louis Vuitton Moet & Hennessy) purchased the brand, though it continues to be called TAG Heuer—even if various very successful reissued classics at first still only bear a simple "Heuer" on their dial.

Founded:
1860 by Edouard Heuer in St.-Imier

Headquarters:
Biel/Bienne

Status:
active without interruption (LVMH today)

Specialty:
chronographs, pilot's chronographs, stopwatches, dashboard clocks and chronographs

Major inventions:
- 1869 patent for keyless winding for pocket watches
- 1887 patent for oscillating pinion in chronographs
- 1930 water-resistant wristwatch case
- 1939 water-resistant wrist chronographs
- 1966 Microtimer measuring a thousandth of a second
- 1969 automatic chronograph (Caliber 11)

Leading models:
Autavia, Carrera, Monaco, German Army chronograph

1945
Chronograph

Case: stainless steel, push-down case back, leather strap, Ø 32 mm
Movement: nickel-plated, column-wheel control of chronograph, manual winding
Remarks: nearly new chronograph with 30-minute counter
Estimated value: $1,820 ➔

1945
Chronograph with Complete Calendar

Case: 14-karat yellow gold, push-down case back, leather strap, Ø 35 mm
Movement: Caliber 72C, nickel-plated, column-wheel control of chronograph, manual winding
Remarks: chronograph with complete calendar
Estimated value: $5,200 ➔

1945
Chronograph with Complete Calendar

Case: stainless steel, screw-down case back, leather strap, Ø 37 mm
Movement: rhodium-plated, column-wheel control of chronograph, manual winding
Remarks: chronograph with complete calendar
Estimated value: $3,380 ➔

1940
Chronograph with Complete Calendar

Case: stainless steel, push-down case back, leather strap, Ø 35 mm
Movement: rhodium-plated, column-wheel control of chronograph, manual winding
Remarks: chronograph with complete calendar
Estimated value: $4,120 ➔

1948
Solunar Tides Watch (Abercrombie & Fitch)

Case: stainless steel, screw-down case back, leather strap, Ø 36 mm
Movement: nickel-plated, manual winding
Remarks: rare men's watch; Solunar model with additional tide display at 6 o'clock; this watch was made by Heuer for Abercrombie & Fitch, London
Estimated value: $1,950 ➔

1945
German Air Force Pilot's Watch

Case: stainless steel in several parts, push-down case back, rotating bezel, leather strap, Ø 38 mm
Movement: nickel-plated, finely finished, column-wheel chronograph control, finely finished steel components, polished screws
Remarks: rare chronograph for the German air force with 30-minute counter and military engraving: "Kampfgeschwader 53, 10. Staffel"; knurled bezel with reference
Estimated value: $3,900 ➔

1950
Chronograph

Case: stainless steel, screw-down case back, leather strap, Ø 35 mm
Movement: nickel-plated, column-wheel control of chronograph, manual winding
Remarks: chronograph with 30-minute counter
Estimated value: $1,560 ➔

1950
Chronograph Complete Calendar Moon Phase

Case: stainless steel, push-down case back, leather strap, Ø 36 mm
Movement: Valjoux Caliber 730, gold-plated, column-wheel control of chronograph, manual winding
Remarks: chronograph with complete calendar and moon phase
Estimated value: $5,850 ➔

HEUER CARRERA

MODERN TIMES

When Jack W. Heuer—freshly crowned majority stockholder of the Heuer company—had finished designing his chronograph, he still lacked a fetching name for the model. What appears self-evident to us nowadays was by no means common in the watch industry in 1963, but times were beginning to change. Objects were given names, and in automobile design, furniture, and architecture, rectilinear shapes had become the norm. This appealed to the young CEO of the venerable old watch brand, so for his new chronograph, he opted for a tidy dial arrangement with straight hands that glowed at night. The totalizer scales were kept understated; the minute scale was banished to the edge of the dial, the flange. This approach was typical of the 1950s, but ultimately also applied to many cars during that era of change.

At the time, Heuer was on a business trip in the United States visiting the Heuer-Léonidas Corporation. He dropped in on the 12-hour Sebring race, which took place in 1963. In the drivers' camp he struck up a conversation with a Mexican family—as it turned out, it was the parents of the two up-and-coming (and very young) Rodríguez brothers, Pedro and Ricardo, to whom Enzo Ferrari had entrusted one of the factory's race cars. Heuer asked the Rodríguezes' mother whether she wasn't worried sick about her sons before the race, but she only waved him off: "This kind of racing circuit is only half as dangerous as the Carrera." What she meant, of course, was the Carrera Panamericana, a grueling race across public roads in Central America that the Mexican president had already prohibited years earlier due to its hazards. Carrera—that was exactly the name Heuer was looking for. And so, in 1963, a legend was born.

1948
Chronograph with Complete Calendar

Case: stainless steel, push-down case back, leather strap, Ø 35 mm
Movement: nickel-plated, column-wheel control of chronograph, manual winding
Remarks: chronograph with complete calendar
Estimated value: $4,160 →

1952
Chronograph

Case: 18-karat yellow gold, push-down case back, leather strap, Ø 35 mm
Movement: nickel-plated, column-wheel control of chronograph, manual winding
Remarks: elegant chronograph
Estimated value: $3,250 →

1960
Chronograph

Case: 14-karat yellow gold, bipartite, screw-down case back, leather strap
Movement: rhodium-plated, column-wheel control of chronograph, fine matte steel chronograph components
Remarks: chronograph with 30-minute and 12-hour counters; printed telemeter (red) and tachymeter (blue) scales
Estimated value: $3,770 →

1960
Chronograph

Case: stainless steel, screw-down case back, leather strap, Ø 37 mm
Movement: Valjoux Caliber 72, nickel-plated, column-wheel control of chronograph, manual winding
Remarks: chronograph with 30-minute and 12-hour counters
Estimated value: $2,340 →

1965
Chronograph Carrera

Case: gold-plated, screw-down case back, leather strap, Ø 36 mm
Movement: Valjoux Caliber 72, nickel-plated, column wheel control of chronograph, manual winding
Remarks: nearly new chronograph with 30-minute and 12-hour counters
Estimated value: $1,950 →

1965
Chronograph Gold

Case: 18-karat yellow gold, screw-down case back, leather strap, Ø 34 mm
Movement: Valjoux Caliber 23R, rhodium-plated, finely polished steel chronograph parts, column-wheel control, glucydur screw balance with Incabloc shock absorber, 17 jewels, manual winding
Remarks: elegant chronograph in almost mint condition
Estimated value: $4,225 →

1965
Chronograph Carrera

Case: stainless steel, screw-down case back, leather strap, Ø 35 mm
Movement: Valjoux Caliber 7730, nickel-plated, manual winding
Remarks: simple sports chronograph, reference number 7753
Estimated value: $2,080 →

1950
Chronograph Complete Calendar Moon Phase

Case: 14-karat gold, push-down steel case back, leather strap, Ø 34 mm
Movement: Valjoux Caliber 88, nickel-plated, column-wheel control of chronograph, 17 jewels, manual winding
Remarks: gold chronograph with complete calendar and the moon phase
Estimated value: $5,850 →

1975
Pilot's Chronograph

Case: stainless steel, screw-down case back, rotating bezel, leather strap, Ø 39 mm
Movement: Valjoux Caliber 7733, nickel-plated, manual winding
Remarks: pilot's chronograph for the Argentinean air force (Fuerza Aerea Argentina) with 30-minute counter
Estimated value: $3,250 ➔

1970
German Army Pilot's Chronograph

Case: stainless steel, screw-down case back, rotating bezel, leather strap, Ø 43 mm
Movement: Valjoux Caliber 230, nickel-plated, column-wheel control of chronograph functions, manual winding
Remarks: Germany army pilot's chronograph with 30-minute counter and flyback function; degree of luminosity printed on dial (3H); number 6645-12463774
Estimated value: $2,600 ➔

1991
German Army Chronograph

Case: stainless steel, tripartite, screw-down case back, rotating bezel, leather strap, Ø 43 mm
Movement: Valjoux Caliber 230, rhodium-plated, column-wheel chronograph control, hacking seconds, polished steel components, 17 jewels
Remarks: pilot's chronograph with 30-minute counter; movement regulated on sidereal time; on case back "6645-12-146-5018"; offered with leather pouch and accessories
Estimated value: $3,900 ➔

1960
Chronograph Autavia Manual Winding

Case: stainless steel, screw-down case back, rotating bezel, Ø 40 mm
Movement: Valjoux Caliber 72, nickel-plated, manual winding
Remarks: sporty stainless steel chronograph
Estimated value: $3,250 ➔

1975
Tides Chronograph

Case: stainless steel, push-down case back, rotating bezel, Ø 41 mm
Movement: Valjoux Caliber 721, rhodium-plated, column-wheel control of chronograph, manual winding
Remarks: large chronograph with additional tide display at 9 o'clock
Estimated value: $3,250 ➔

1965
Chronograph Autavia GMT

Case: stainless steel, push-down case back, rotating bezel with 24-hour indication, leather strap, Ø 41 mm
Movement: Valjoux Caliber 724, rhodium-plated, finely polished steel chronograph parts, column-wheel control, glucydur balance, manual winding
Remarks: rare chronograph with additional 24-hour hand showing a second time zone
Estimated value: $4,745 ➔

1970
Chronograph Autavia

Case: stainless steel, tripartite, screw-down case back, rotating bezel, leather strap, Ø 41 mm
Movement: Valjoux Caliber 7733, nickel-plated; slideway control of chronograph; mirror-polished chronograph components
Remarks: large chronograph with 30-minute counter whose 5-minute intervals are underscored; additional minute scale on the rotating bezel
Estimated value: $2,080 ➔

1975
Chronograph Montreal

Case: stainless steel, screw-down case back, stainless steel link bracelet, 42 x 48 mm
Movement: Caliber 12, red gold-plated, micro rotor, automatic winding
Remarks: heavy automatic chronograph with 30-minute and 12-hour counters; date window at 6 o'clock
Estimated value: $1,560 ➔

1975

Chronograph Autavia

Case: stainless steel, screw-down case back, rotating bezel, leather strap, 42 x 48 mm
Movement: Caliber 12, gold-plated, micro rotor, automatic winding
Remarks: heavy automatic chronograph with 30-minute and 12-hour counters; date window at 6 o'clock
Estimated value: $2,080 →

1975

Chronograph Autavia GMT

Case: stainless steel, screw-down case back, rotating bezel, stainless steel link bracelet, Ø 42 mm
Movement: Caliber 11, gold-plated, micro rotor, 17 jewels, automatic winding
Remarks: automatic chronograph with 30-minute and 12-hour counters; additional 24-hour display
Estimated value: $3,380 ↗

1975

Chronograph Carrera

Case: stainless steel, screw-down case back, stainless steel link bracelet, Ø 39 mm
Movement: Caliber 12, gold-plated, micro rotor, automatic
Remarks: heavy automatic chronograph with 30-minute and 12-hour counters
Estimated value: $1,950 →

1970

Chronograph Calculator

Case: stainless steel, screw-down case back, rotating bezel, leather strap, 36 x 43 mm
Movement: Caliber 12, gold-plated, micro rotor, automatic winding
Remarks: automatic chronograph with slide rule function integrated into the rotating bezel
Estimated value: $2,600 →

1975

Chronograph Carrera

Case: gold-plated, screw-down case back, leather strap, Ø 38 mm
Movement: Caliber 12, gold-plated, micro rotor, automatic winding
Remarks: automatic chronograph with 30-minute and 12-hour counters
Estimated value: $1,300 →

1968

Chronograph Camaro

Case: stainless steel, screw-down case back, leather strap, 37 x 38 mm
Movement: Valjoux Caliber 72, rhodium-plated, finely polished steel chronograph parts, column-wheel control, glucydur balance, 17 jewels, manual winding
Remarks: sporty, cushion-shaped chronograph
Estimated value: $1,950 →

1975

Chronograph Monaco Manual Winding

Case: stainless steel, push-down case back, leather strap, 40 x 45 mm
Movement: nickel-plated, manual winding
Remarks: chronograph with 30-minute and 12-hour counters
Estimated value: $7,150 →

1975

Chronograph Monaco

Case: stainless steel, leather strap, 40 x 45 mm
Movement: Caliber 12, gold-plated, micro rotor, automatic winding
Remarks: heavy automatic chronograph with 30-minute counter and date window
Estimated value: $6,500 →

IWC

When twenty-seven-year-old American engineer Florentine Ariosto Jones, hailing from Boston, founded International Watch Company in Switzerland, he didn't do it simply because Switzerland had become a land of low wages in comparison to the United States in the mid-nineteenth century. The American method of manufacturing watches with highly automated processes had been reaching the boundaries of its capacity at home, and on searching out a new production location, he found that Switzerland was perfect for his goals with its horological know-how and low wages.

In the Swiss Jura region, where most watchmakers made their homes, no one wanted to listen to anything about automated machine production—a method that required only a few workers. But Jones received an offer from the city of Schaffhausen, and one potential business partner was a certain Johann Heinrich Moser. This man had invented a new system for industrially harnessing the power of the Rhine's flow. Gear wheels and transmission ropes were powered by waterwheels, which in turn conveyed their energy to machines. Thanks to Moser's invention, Jones's new company could receive its energy practically for free. Jones saw the wisdom of this, and Moser became one of the founders of International Watch Company in 1869. Just one year later, IWC was producing 10,000 watches—all of them for the American market.

Exports began to slow down in 1874, however, because the United States had started levying a 25 percent protective duty on foreign goods. The powers that be turned IWC into a joint stock company. Jones returned to Boston, and IWC was bought by the Schaffhausen Handelsbank.

Four years later, the entrepreneurial Rauschenbach family took over the helm. They opened new markets for IWC—most especially in neighboring Germany where the watches were in high demand thanks to their excellent quality.

The company soon recognized the importance of aviation and offered pilots precision watches that they could also wear over their thick jackets. The tradition of IWC pilot's watches, which began here, is something that the Schaffhausen-based company has kept alive to this day.

It would, however, be unfair to confine IWC to being a manufacturer of pilot's watches alone. IWC's movements were known as exceptionally solid in design and technically well thought out—and not only in collector's circles. The simple three-handed models of the Schaffhausen *manufacture* were also famous, embodying the simple understatement of these high-quality watches. IWC also walked down its own path in terms of cases. The Ingenieur was launched in 1954: a soft iron core within its case protected the movement from the influence of magnetic fields which would normally impact rate precision. In the 1970s, IWC was sold to the VDO/Mannesmann concern together with Jaeger-LeCoultre, finally ending up under the umbrella of the Richemont Group in the new millennium.

Founded:
1868 by Florentine Ariosto Jones in Schaffhausen

Headquarters:
Schaffhausen

Status:
active without interruption (today Richemont Luxury Group)

Specialty:
pilot's watches, antimagnetic watches, chronographs, diving watches

Major inventions:
- 1950 Pellaton automatic winding with cam
- 1969 contribution to the development of the Beta 21
- 1980 first titanium case

Leading models:
Ingenieur, Portuguese, Aquatimer, Yacht Club, Pilot's Watch

1920
Men's Watch

Case: 18-karat yellow gold, tripartite, push-down case back, leather strap, 23 x 35 mm
Movement: gold-plated, frosted finish, jewels set in chatons
Remarks: extremely rare early men's watch in a gold Art Deco case; obvious asymmetrical arrangement of the movement, which was pushed toward the 6 o'clock position, while the upper part of the case is decorated with a black Art Deco motif
Estimated value: $3,900 →

1927
Men's Watch

Case: 14-karat yellow gold, bipartite, push-down case back, leather strap, 24 x 35 mm
Movement: Caliber 84, gold-plated, frosted finish, jewels set in chatons, polished screws
Remarks: fine men's watch in a rectangular Art Deco case; the slanted surfaces of the case are finely guilloché; this watch was offered with an IWC buckle
Estimated value: $3,250 →

1915
Military Watch

Case: silver, locking cover over the crystal, hinged back cover, leather strap, Ø 35 mm
Movement: grained gold plating, jewels set in gold chatons, compensation screw balance, manual winding
Remarks: extremely rare early "trench watch" with a protective cover
Estimated value: $4,550 ↘

1937
Men's Watch

Case: stainless steel, fluted edges, push-down case back, leather strap, 24 x 40 mm
Movement: rhodium-plated, jewels set in chatons, decoration, glucydur screw balance, manual winding
Estimated value: $3,900 ↘

ca. 1938
Men's Watch for Türler

Case: stainless steel, push-down case back, leather strap, 30 x 21 mm
Movement: IWC Caliber 87, rhodium-plated, shaped case, fausses côtes decoration, jewels set in chatons, polished screws, manual winding
Remarks: rectangular men's watch in double signature IWC/Türler; this watch was made for the jeweler Türler
Estimated value: $2,600 →

1941
Men's Watch

Case: 14-karat yellow gold, push-down case back, leather strap, Ø 36 mm
Movement: IWC Caliber 83, rhodium-plated, côtes de Genève, jewels set in chatons, manual winding
Remarks: men's watch in yellow gold case
Estimated value: $1,560 →

1941
Men's Watch

Case: stainless steel, push-down case back, leather strap, Ø 35 mm
Movement: IWC Caliber 83, nickel-plated, côtes de Genève, manual winding
Remarks: early IWC men's watch with subsidiary seconds
Estimated value: $1,170 →

1943
Men's Watch

Case: stainless steel, push-down case back, leather strap, Ø 35 mm
Movement: IWC Caliber 83, rhodium-plated, côtes de Genève, manual winding
Remarks: simple men's watch with subsidiary seconds; applied hour numerals are highlighted with a light-colored ring
Estimated value: $1,300 →

1941
Men's Watch

Case: stainless steel, push-down case back, leather strap, Ø 35 mm
Movement: IWC Caliber 83, nickel-plated, côtes de Genève, manual winding
Remarks: rare, early IWC men's watch
Estimated value: $1,300 ➔

1954
Men's Watch

Case: 18-karat yellow gold, push-down case back, leather strap, Ø 37 mm
Movement: IWC Caliber 89, rhodium-plated, côtes de Genève, jewels set in chatons, manual winding
Remarks: elegant men's watch in yellow gold case with highly noticeable shaped strap lugs
Estimated value: $1,560 ➔

1956
Men's Watch

Case: 18-karat yellow gold, push-down case back, leather strap, Ø 33 mm
Movement: IWC Caliber 89, rhodium-plated, côtes de Genève, jewels set in chatons, manual winding
Remarks: simple men's watch with strongly accentuated strap lugs
Estimated value: $1,170 ➔

1952
Men's Watch

Case: 18-karat yellow gold, push-down case back, leather strap, Ø 36 mm
Movement: IWC Caliber 89, rhodium-plated, côtes de Genève, jewels set in chatons, manual winding
Remarks: elegant men's watch in yellow gold case
Estimated value: $1,560 ➔

1950
Men's Watch with Enamel Dial

Case: 18-karat rose gold, push-down case back, leather strap, Ø 37 mm
Movement: rhodium-plated, fausses côtes decoration, jewels set in gold chatons, glucydur screw balance, manual winding
Remarks: unique cloisonné enamel dial after Charles Poluzzi
Estimated value: $9,100 ➔

1949
Men's Watch with Enamel Dial

Case: 18-karat yellow gold, push-down case back, leather strap, Ø 36 mm
Movement: Caliber 89, rhodium-plated, fausses côtes decoration, jewels set in gold chatons, glucydur screw balance, manual winding
Remarks: unique cloisonné enamel dial with the Israeli flag, made a year after the founding of the state of Israel
Estimated value: $13,200 ➔

1928
Men's Watch

Case: stainless steel, push-down steel case back, leather strap, Ø 35 mm
Movement: gold-plated, frosted finish, manual winding
Remarks: very early wristwatch in what was for the time an exclusive stainless steel case
Estimated value: $780 ➔

1980
Skeleton Watch

Case: 18-karat yellow gold, push-down transparent case back, leather strap, Ø 34 mm
Movement: Piguet Caliber, gold rotor, completely hand-skeletonized and decorated movement, glucydur balance, manual winding
Remarks: very fine skeletonized wristwatch, produced in small series
Estimated value: $6,500 ➚

1943
Mark IX

Case: stainless steel, push-down case back, leather strap, Ø 37 mm
Movement: rhodium-plated, jewels set in chatons, glucydur screw balance, manual winding
Remarks: extremely rare "special watch for flyers" (only 430 pieces registered); with excerpt from the IWC registry
Estimated value: $23,500 →

1940
Big Pilot's Watch

Reference number: 431
Case: stainless steel, matte, protective cap on movement, push-down case back, leather strap, Ø 55 mm
Movement: Caliber 52T 19''' H6 S.C., gold-plated, frosted finish, jewels set in chatons, indirect sweep second with a stop system, compensation screw balance, manual winding
Remarks: extremely rare large flyer's watch of the German Luftwaffe with original box and excerpt from the IWC register
Estimated value: $26,000 ↘

1944
Mark X

Case: stainless steel, bipartite, push-down case back, leather strap, Ø 35 mm
Movement: Caliber 83, rhodium-plated, jewels set in chatons, fausses côtes decoration, polished screws
Remarks: early pilot's watch for the British Royal Air Force with military case back engraving "W.W.W.M17417"; this watch was offered with an IWC buckle
Estimated value: $5,200 →

1944
Mark X

Case: stainless steel, push-down case back, leather strap, Ø 35 mm
Movement: Caliber IWC 83, nickel-plated, côtes de Genève, jewels set in chatons, manual winding
Remarks: Mark X pilot's watch for the Royal Air Force with case back engraving "W.W.W.M14177"
Estimated value: $5,200 →

1951
Mark XI (British Armed Forces)

Case: stainless steel, screw-down case back, leather strap, Ø 36 mm
Movement: Caliber IWC 89, nickel-plated, côtes de Genève, manual winding
Remarks: pilot's watch for the Royal Air Force; this watch has an additional antimagnetic soft iron core
Estimated value: $5,200 →

1951
Mark XI BOAC

Case: stainless steel, screw-down case back, leather strap, Ø 36 mm
Movement: Caliber IWC 89, nickel-plated, côtes de Genève, manual winding
Remarks: extremely rare pilot's watch for the Royal Air Force; after World War II, this watch was used by the British Overseas Aircraft Cooperation (BOAC); the BOAC logo is engraved on the case back
Estimated value: $5,980 →

1963
Mark XI (civilian model)

Case: stainless steel, screw-down case back, leather strap, Ø 36 mm
Movement: IWC Caliber 89, nickel-plated, côtes de Genève, manual winding
Remarks: this watch was offered with its original box and papers
Estimated value: $5,200 ↗

1969
Ingenieur

Reference number: 866A
Case: stainless steel, screw-down case back, soft iron shield, leather strap, Ø 37 mm
Movement: Caliber 854, rhodium-plated, decoration, glucydur screw balance, automatic winding
Remarks: extremely rare Ingenieur in a flyer's version, used for testing purposes by Germany's Air Force
Estimated value: $19,500 ↗

1971
Aquatimer Automatic

Case: stainless steel, bipartite, screw-down case back, leather strap, Ø 37 mm
Movement: Caliber 8541B, rhodium-plated, fausses côtes decoration, automatic winding, 21 jewels
Remarks: nearly new diver's watch; the dive time is set on the rotating flange using the second crown
Estimated value: $7,800 →

1971
30 ATM Aquatimer Automatic

Case: stainless steel, screw-down case back, rotating inner diver's bezel, leather strap, Ø 40 mm
Movement: IWC Caliber 8541B, nickel-plated, automatic winding
Remarks: water-resistant to 300 meters; date window at 3 o'clock
Estimated value: $5,850 →

1985
Porsche Design

Case: aluminum, black anodized, case back opens to reveal a compass and mirror, metal bracelet, Ø 39 mm
Movement: rhodium-plated, glucydur balance, automatic winding
Remarks: rare early Porsche Design watch with compass
Estimated value: $3,250 →

1995
Porsche Design Ocean

Case: titanium, screw-down case back, rotating bezel with markers, titanium bracelet, Ø 42 mm
Movement: gold-plated, polished, glucydur balance, 22 jewels, automatic winding
Remarks: diver's watch, water-resistant to 200 atm, with additional gray Velcro tape and accessories
Estimated value: $3,650 ↗

1936
Portuguese

Case: stainless steel, push-down case back, leather strap, Ø 42 mm
Movement: Caliber 98, rhodium-plated, decoration, jewels set in gold chatons, compensation screw balance, manual winding
Remarks: significant men's wristwatch, so-called "Original Portuguese," with the legendary pocket watch Caliber 98; with excerpt from the IWC registry
Estimated value: $26,000 →

1975
Portuguese

Case: stainless steel, push-down caseback, Ø 42 mm
Movement: IWC Caliber 982, rhodium-plated, côtes de Genève, jewels set in chatons, manual winding
Remarks: first new edition of the legendary Portuguese series; this watch was offered in its original box with papers
Estimated value: $39,000 ↗

1993
New Portuguese

Case: stainless steel, push-down transparent case back, leather strap, Ø 42 mm
Movement: Caliber 9828, rhodium-plated, fausses côtes decoration, jewels set in gold chatons, glucydur balance with shock protection, 19 jewels, regulated in 5 positions, manual winding
Remarks: from a limited edition of 1,000 pieces for the relaunch of the 1993 model
Estimated value: $15,600 →

1953
Automatic

Case: 18-karat yellow gold, push-down case back, leather strap, 32 x 41 mm
Movement: Caliber 8521, rhodium-plated, côtes de Genève, glucydur screw balance with Incabloc shock protection, automatic winding
Remarks: fine automatic men's watch with sweep seconds and date
Estimated value: $3,250 →

1960
Ingenieur

Case: 18-karat yellow gold, tripartite, screw-down case back, leather strap, Ø 36 mm
Movement: Caliber 8531, rhodium-plated, fausses côtes decoration, automatic winding, index fine adjustment, 21 jewels
Remarks: elegant gold men's watch
Estimated value: $5,850 ➔

1963
Ingenieur

Case: stainless steel, tripartite, screw-down case back, leather strap, Ø 36 mm
Movement: Caliber 8531, rhodium-plated, fausses côtes decoration, automatic winding, index fine adjustment, 21 jewels
Remarks: sporty Ingenieur model
Estimated value: $5,200 ➔

1970
Men's Watch

Case: 18-karat yellow gold, monocoque, Ø 34 mm
Movement: rhodium-plated, fausses côtes decoration
Remarks: elegant men's watch; this timepiece was offered in its original box
Estimated value: $1,560 ➔

1953
Automatic

Case: 18-karat yellow gold, push-down case back, leather strap, Ø 33 mm
Movement: IWC Caliber 8521, nickel-plated, côtes de Genève, 21 jewels, automatic
Remarks: automatic wristwatch with the famous IWC eccentric automatic winding; date window at 3 o'clock
Estimated value: $1,950 ➔

1963
Automatic

Case: stainless steel, push-down case back, stainless steel link bracelet, Ø 35 mm
Movement: IWC Caliber 8531, 21 jewels, automatic winding
Remarks: simple stainless steel men's watch; date window at 3 o'clock
Estimated value: $1,820 ➔

1969
Automatic

Case: stainless steel, push-down case back, leather strap, Ø 35 mm
Movement: Caliber IWC 8541, rhodium-plated, with côtes de Genève, 23 jewels, automatic
Remarks: this IWC movement was regulated in 5 positions; date window at 3 o'clock
Estimated value: $1,690 ➔

1966
Automatic

Case: 18-karat white gold, push-down case back, 18-karat white gold Milanaise bracelet, Ø 35 mm
Movement: Caliber 8541, rhodium-plated,polished, polished screws, eccentric fine adjustment, automatic winding
Remarks: elegant men's watch in white gold case with white gold Milanaise bracelet
Estimated value: $5,200 ➔

1959
Automatic

Case: 18-karat yellow gold, push-down case back, gold Milanaise bracelet, Ø 35 mm
Movement: Caliber 8531, rhodium-plated, automatic winding, index fine adjustment, 21 jewels
Remarks: fine automatic men's watch with date display
Estimated value: $3,640 ➔

1961
Automatic De Luxe

Case: yellow gold, push-down case back, gold link bracelet, Ø 35 mm
Movement: IWC Caliber 8531, nickel-plated, côtes de Genève, 21 jewels, automatic winding
Remarks: rarely worn heavy gold watch with original IWC gold bracelet and original certificate; date window at 3 o'clock
Estimated value: $7,800 ↗

1964
Yacht Club

Case: stainless steel, bipartite, screw-down case back, leather strap, Ø 31 mm
Movement: Caliber 8541, côtes de Genève, polished screws, index fine adjustment
Remarks: automatic Yacht Club in a small case
Estimated value: $1,950 →

1969
Yacht Club

Case: stainless steel, screw-down case back, leather strap, Ø 36 mm
Movement: IWC Caliber 8541B, nickel-plated, date, 23 jewels, automatic winding
Remarks: sporty automatic watch in modern stainless steel case
Estimated value: $2,080 →

1972
Yacht Club Automatic

Case: stainless steel, screw-down case back, leather strap, Ø 36 mm
Movement: Caliber 8541B, rhodium-plated, decoration, 23 jewels, regulated in 5 positions, automatic winding
Remarks: automatic men's watch with sweep seconds and date
Estimated value: $3,900 →

1970
Yacht Club Automatic

Case: 18-karat yellow gold, screw-down case back, leather strap, Ø 35 mm
Movement: rhodium-plated, côtes de Genève, glucydur balance, automatic winding
Remarks: gift watch of the royal house of Oman
Estimated value: $4,550 →

1976
Yacht Club II

Case: stainless steel, screw-down case back, stainless steel link bracelet, Ø 38 mm
Movement: IWC Caliber 3254, rhodium-plated, côtes de Genève, automatic winding
Remarks: the octagonal bezel of the Yacht Club II is remarkable
Estimated value: $2,340 →

1959
Ingenieur

Case: 18-karat red gold, tripartite, screw-down case back, leather strap, Ø 36 mm
Movement: Caliber 8531, rhodium-plated, fausses côtes decoration, automatic winding, index fine regulation, 21 jewels
Remarks: elegant gold Ingenieur model
Estimated value: $7,150 →

1963
Ingenieur

Reference number: 666A
Case: 18-karat yellow gold, tripartite, screw-down case back, leather strap, Ø 36 mm
Movement: Caliber 8531, rhodium-plated, fausses côtes decoration, automatic winding, index fine adjustment, 17 jewels
Remarks: gold Ingenieur; sold to Blech in Copenhagen on October 26, 1963; the watch was offered with a certificate from the IWC master registry
Estimated value: $7,150 →

1967
Ingenieur

Reference number: 866
Case: stainless steel, screw-down case back, soft iron shield, leather strap, Ø 37 mm
Movement: Caliber 854, rhodium-plated, decoration, glucydur screw balance, automatic winding
Remarks: extremely rare antimagnetic men's wristwatch with an automatic movement; only example of the famed Ref. 866 without a date display
Estimated value: $13,100 ➔

1973
Ingenieur SL Jumbo

Case: 18-karat yellow gold, screw-down case back, 18-karat gold link bracelet, Ø 40 mm
Movement: Caliber IWC 8541B, rhodium-plated, côtes de Genève, automatic winding
Remarks: especially heavy (260 grams) gold watch; like all Ingenieurs, this one is well protected from magnetic fields
Estimated value: $16,900 ➚

1975
Da Vinci Quartz Electronic

Case: 18-karat yellow gold, screw-down case back, 18-karat gold link chain, 38 x 44 mm
Movement: Caliber CEH Beta 21, quartz
Remarks: heavy gold watch in a hexagonal case, equipped with one of the first quartz movements
Estimated value: $4,550 ➔

1996
Da Vinci Rattrapante

Reference number: 3751
Case: 18-karat yellow gold, push-down case back, leather strap, Ø 39 mm
Movement: Caliber 79251, gold-plated, polished steel chronograph parts, automatic
Remarks: elegant rattrapante chronograph with perpetual calendar and moon phase, four-digit year display
Estimated value: $19,500 ➔

1978
Ingenieur SL Automatic

Reference number: 1832
Case: stainless steel, tripartite, screw-down case back, stainless steel link bracelet, Ø 40 mm
Movement: Caliber 8541, rhodium-plated, decorated, polished screws
Remarks: large, heavy Ingenieur model with stainless steel link bracelet; this watch was offered in its original box, with original certificate and guarantee and a certificate from the IWC master registry
Estimated value: $11,700 ➔

1983
Chronograph

Reference number: 3710
Case: 18-karat yellow gold, screw-down case back, leather strap, Ø 36 mm
Movement: ETA Caliber 795, gold-plated fausses côtes decoration, finely polished steel chronograph parts, column-wheel control, glucydur balance with shock protection, manual winding
Remarks: elegant chronograph with a 30-minute and 12-hour counter, moon phase display, from a limited edition of 150 pieces
Estimated value: $10,400 ➔

1988
Da Vinci Chronograph

Reference number: 3750
Case: stainless steel, push-down case back, leather strap, Ø 38 mm
Movement: Caliber 7906, gold-plated, polished steel chronograph parts, automatic
Remarks: elegant chronograph with perpetual calendar and moon phase, four-digit year display
Estimated value: $9,750 ➔

1991
Grande Complication

Reference number: 3770
Case: platinum, screw-down case back, leather strap, Ø 42 mm
Movement: Caliber 79091, gold-plated, automatic
Remarks: large watch with complications, with minute repeater, chronograph and perpetual calendar with moon phase; limited to 50 pieces
Estimated value: $91,000 ➔

JAEGER-LECOULTRE

The name of this company should actually be flipped, for the history of Jaeger-LeCoultre began with Antoine LeCoultre, born in 1803 in Le Sentier, in the Vallée de Joux, Switzerland's "watch valley," where the roots of his family extend back to the sixteenth century. At the age of thirty he founded a factory for movement components in his hometown, on the very spot where Jaeger-LeCoultre continues to operate today.

Antoine LeCoultre was a very imaginative man. In 1844, he invented the millionometer, an instrument that made it possible to take measurements with the precision of a thousandth of a millimeter (or a millionth of a meter—hence the name). Even the double function of a crown for winding and setting, which is so natural to us today, is one of LeCoultre's inventions, and finally made key winding superfluous. LeCoultre registered many patents, and after his death, the *manufacture* LeCoultre & Compagnie specialized in producing fine, complicated watch movements.

The second part of this company's history began in 1925 with the addition of Alsatian engineer Edmond Jaeger (born in 1850), who had previously supplied both Cartier and the French navy, among others, with movements. The *manufacture's* double name gave him the opportunity to develop and produce not only movements, but also entire wristwatches. The Duoplan was created in 1926 with a bipartite baguette movement that could be easily exchanged in just a few minutes should it become damaged. The famous Atmos clock followed in 1928, a timepiece that gets its energy from variations in temperature in its environment. To this day, it remains the official Swiss gift to foreign state guests. In 1929, Caliber 101 was added to the repertoire, the smallest mechanical movement ever serially manufactured. And, finally, in 1931 the famous Reverso model followed, whose easily rotatable case served to protect the sensitive crystal on the front. According to legend, the inspiration for this exceptional case design came from British polo players stationed in India.

Jaeger-LeCoultre soon enjoyed an excellent reputation as a *manufacture* to watch. Its specialties, then and now, counted calendar watches with moon phases, perpetual calendars, and alarm watches among them. In 1948, after the death of Jacques-David LeCoultre, the company was purchased by SAPIC Holding. It then passed through a number of hands until it finally found a home within the VDO/Mannesmann concern at the end of the 1970s. Just about twenty years later, that company sold its entire watch holdings—called Les Manufactures Horlogères, including Jaeger-LeCoultre, IWC, and A. Lange & Söhne—to the Richemont Group.

Founded:
1833 by Antoine LeCoultre in Le Sentier

Headquarters:
Le Sentier

Status:
active without interruption (today Richemont Luxury Group)

Specialty:
all types of complications, including repeaters and tourbillons, ultra-thin calibers, chronographs, wristwatch alarms

Major inventions:
- 1844 invention of the millionometer (measurement tool)
- 1847 crown winder with lever
- 1928 Atmos, driven by temperature and atmospheric pressure
- 1929 Caliber 101, to this day, the smallest mechanical caliber
- 1931 Reverso, with a reversible case

Leading models:
Reverso, Futurematic, Memovox

1935
Reverso

Case: stainless steel, push-down case back, leather strap, 23 x 38 mm
Movement: gold-plated, côtes des Genève, manual winding
Remarks: one of the first Reverso models in the famous reversible case with sweep seconds
Estimated value: $5,200 ➔

1935
Reverso Sawaiman Guards

Case: stainless steel, push-down case back, leather strap
Movement: rhodium-plated, fausses côtes decoration, 15 jewels, manual winding
Remarks: extremely rare Reverso model with engraved and enameled coat of arms of the Sawaiman Guards, the personal body guards of Maharaja Sawaiman of Jaipur
Estimated value: $7,800 ➔

1940
Reverso

Case: 9-karat yellow gold, push-down case back, leather strap
Movement: rhodium-plated, fausses côtes decoration, manual winding
Remarks: early rare Reverso model with sweep seconds in the famous reversible case; this watch has a hand-engraved coat of arms on the case back
Estimated value: $5,200 ➔

1935
Reverso Maharani

Case: 18-karat yellow gold, push-down case back, leather strap
Movement: rhodium-plated, fausses côtes decoration, manual winding
Remarks: only few pieces of the Reverso Maharani were produced, and these were presumed missing up to now; this watch was made for the Maharaja of Karputala, who had his wife perpetuated on its case back thanks to a lavish enamel technique
Estimated value: $91,000 ➔

1950
Reverso Luxe (Meister, Rio de Janeiro)

Case: stainless steel, leather strap, 23 x 38 mm
Movement: rhodium-plated, jewels set in chatons, shaped movement, manual winding
Remarks: rare Reverso model with sweep seconds; dial printed with "Meister, Rio de Janeiro"
Estimated value: $5,200 ➔

1950
Calendar Watch

Case: stainless steel, push-down case back, leather strap, 24 x 40 mm
Movement: shaped movement, nickel-plated, côtes de Genève, manual winding
Remarks: rectangular men's watch with calendar, sweep date hand, weekday in the window at 12 o'clock, and subsidiary seconds
Estimated value: $2,600 ➔

1945
Calendar Watch

Case: gold-plated, push-down case back, leather strap, 24 x 40 mm
Movement: Caliber 806/AW, rhodium-plated, côtes de Genève, manual winding
Remarks: rare rectangular men's watch with complete calendar and moon phase; sweep date hand; weekday and month in window display
Estimated value: $3,250 ➔

1950
Calendar Watch

Case: 18-karat yellow gold, push-down case back, leather strap, 24 x 40 mm
Movement: shaped movement, nickel-plated, côtes de Genève, manual winding
Remarks: rectangular men's watch with calendar, sweep date hand, weekday in window display at 12 o'clock, subsidiary seconds
Estimated value: $3,900 ➔

1945
Men's Watch

Case: 10-karat rolled gold, push-down steel case back, leather strap, 27 x 39 mm
Movement: Caliber 438/4CW, nickel-plated, côtes de Genève, manual winding
Remarks: rare men's watch in an unusual design case
Estimated value: $650 ➔

1945
Men's Watch

Case: stainless steel, push-down case back, leather strap, 28 x 36 mm
Movement: gold-plated, côtes de Genève, manual winding
Remarks: men's wristwatch in rectangular case
Estimated value: $650 ➔

1940
Futurematic

Case: 14-karat yellow gold, push-down case back, leather strap, Ø 35 mm
Movement: Caliber 497, rhodium-plated, côtes de Genève, automatic winding
Remarks: automatic men's watch with hammer automatic and power reserve indicator at 9 o'clock; the crown is hidden on the case back to visually underscore that the watch is self-winding when worn; this watch is only signed "LeCoultre"
Estimated value: $1,820 ➔

1950
Futurematic

Case: gold-plated, push-down case back, leather strap, Ø 35 mm
Movement: Caliber 497, rhodium-plated, côtes de Genève, automatic winding
Remarks: automatic men's watch with hammer automatic and power reserve indicator at 9 o'clock; the crown is hidden on the case back to visually underscore that the watch is self-winding when worn
Estimated value: $1,300 ➔

1950
Men's Watch with Digital Display

Case: 18-karat rose gold, push-down case back, leather strap, 27 x 35 mm
Movement: Caliber 480/CW, nickel-plated, côtes de Genève, manual winding
Remarks: rare men's watch with digital display of the time
Estimated value: $2,600 ➔

1940
Men's Watch

Case: stainless steel, push-down case back, leather strap, Ø 30 mm
Movement: shaped movement, nickel-plated, côtes de Genève, manual winding
Remarks: rare wristwatch with calendar, sweep date hand, weekday in window display at 12 o'clock
Estimated value: $1,560 ➔

1945
Automatic with Power Reserve Indicator

Case: gold-plated, screw-down case back, leather strap, Ø 34 mm
Movement: Caliber 481, rhodium-plated, côtes de Genève, automatic winding
Remarks: elegant automatic men's watch (hammer automatic), with power reserve indicator at 12 o'clock
Estimated value: $910 ➔

1949
Calendar Watch

Case: stainless steel, push-down case back, leather strap, Ø 35 mm
Movement: Caliber 464/A, gold-plated, côtes de Genève, manual winding
Remarks: rare steel men's watch with calendar; sweep date hand; window display of weekday and month
Estimated value: $5,200 ➔

1950
Calendar Watch

Case: 18-karat red gold, push-down case back, leather strap, Ø 35 mm
Movement: Caliber 494/A, gold-plated, côtes de Genève, manual winding
Remarks: elegant red gold men's watch with complete calendar and moon phase; sweep date hand; window display of weekday and month; moon phase in small subsidiary dial at 6 o'clock
Estimated value: $5,200 ➔

1950
Semi-Mystérieuse

Case: 18-karat yellow gold, push-down case back, leather strap, Ø 32 mm
Movement: rhodium-plated, côtes de Genève, manual winding
Remarks: rare watch with semi-mysterious time display; the crown is hidden on the case back; this watch was offered in its original box
Estimated value: $3,900 ➔

1950
Mystérieuse

Case: 14-karat white gold, push-down case back, leather strap, Ø 33 mm
Movement: Caliber 918/2, nickel-plated, manual winding
Remarks: rare white gold men's watch with diamond-set dial; two diamonds serve as the hands, which appear to move freely
Estimated value: $3,250 ➔

1980
Chronometer Geomatic

Case: stainless steel, screw-down case back, leather strap, Ø 36 mm
Movement: Caliber K881G, nickel-plated, côtes de Genève, 23 jewels, regulated in 6 positions, automatic winding
Remarks: nearly new steel chronometer with date window at 3 o'clock; officially certified chronometer; the movement was regulated in 6 positions; this watch was offered in its original box with papers
Estimated value: $3,900 ➔

1950
Chronograph

Case: 18-karat yellow gold, screw-down case back, leather strap, Ø 35 mm
Movement: Valjoux Caliber 72, rhodium-plated, column-wheel control of chronograph functions, manual winding
Remarks: chronograph with 30-minute and 12-hour counters; this watch was offered in its box with a LeCoultre sales tag and papers
Estimated value: $3,900 ➔

1950
Chronograph

Case: stainless steel, screw-down case back, leather strap, Ø 35 mm
Movement: Valjoux Caliber 72, nickel-plated, column wheel control of chronograph functions, manual winding
Remarks: simple chronograph with 30 minute and 12-hour counters; only signed "LeCoultre"
Estimated value: $1,950 ➔

1970
Chronograph

Case: stainless steel, screw-down case back, rotating bezel, leather strap, Ø 40 mm
Movement: Valjoux Caliber 72, rhodium-plated, column-wheel control of chronograph, manual winding
Remarks: sporty chronograph with 30-minute and 12-hour counters in heavy steel case with a rotating bezel; only signed "LeCoultre"
Estimated value: $3,250 ➔

1960
Chronograph

Case: stainless steel, screw-down case back, leather strap, Ø 35 mm
Movement: Valjoux Caliber 72, rhodium-plated, column-wheel control of chronograph, manual winding
Remarks: simple chronograph with 30-minute and 12-hour counters; only signed "LeCoultre"
Estimated value: $1,820 ➔

1950
Chronograph
Case: 18-karat yellow gold, push-down case back, leather strap, Ø 36 mm
Movement: Caliber 285, nickel-plated, column-wheel control of chronograph, manual winding
Remarks: rare simple chronograph with 45-minute counter
Estimated value: $3,250 ➔

1951
Memovox
Case: stainless steel, push-down case back, leather strap, Ø 35 mm
Movement: Caliber 489/1, rhodium-plated, manual winding
Remarks: very early Memovox with alarm function
Estimated value: $1,950 ➔

1951
Memovox
Case: stainless steel, push-down case back, leather strap, Ø 35 mm
Movement: Caliber 489/1, rhodium-plated, manual winding
Remarks: very early alarm wristwatch; a reference marker on the inner disk displays the alarm time
Estimated value: $1,560 ➔

1960
Memovox
Case: yellow gold, push-down case back, leather strap, Ø 35 mm
Movement: Caliber 814, rhodium-plated, manual winding
Remarks: rare parking meter wristwatch alarm; to conveniently be reminded of the parking meter running out, it is shown in half-hour increments on the inner disk
Estimated value: $2,080 ➔

1955
Memovox
Case: 18-karat yellow gold, leather strap, Ø 35 mm
Movement: Caliber P489/1, nickel-plated, manual winding
Remarks: alarm wristwatch in gold case; a reference marker displays the set alarm time on the inner disk
Estimated value: $1,950 ➔

1960
Memovox
Case: stainless steel, 18-karat yellow gold bezel, push-down case back, leather strap, Ø 35 mm
Movement: rhodium-plated, manual winding
Remarks: nearly unworn alarm wristwatch in unusual two-tone case with fluted bezel
Estimated value: $2,600 ➔

1960
Memovox
Case: gold-plated, push-down case back, leather strap, Ø 34 mm
Movement: Caliber 814, rhodium-plated, manual winding
Remarks: elegant alarm wristwatch with a company logo on the dial
Estimated value: $1,300 ➔

1957
Memovox Automatic
Case: gold-plated, push-down case back, leather strap, Ø 38 mm
Movement: Caliber 825, nickel-plated, côtes de Genève, automatic winding
Remarks: alarm wristwatch with hammer automatic movement and date window at 3 o'clock; a reference marker displays the set alarm time on the inner disk
Estimated value: $1,950 ➔

1967
Memovox Automatic

Case: stainless steel, screw-down case back, leather strap, Ø 37 mm
Movement: Caliber 916, rhodium-plated, automatic winding
Remarks: discreet alarm wristwatch with automatic movement
Estimated value: $1,950 ➔

1958
Memovox Automatic

Case: 18-karat yellow gold, screw-down case back, leather strap, Ø 37 mm
Movement: Caliber 825, rhodium-plated, automatic winding
Remarks: elegant alarm wristwatch with the first automatic alarm caliber (hammer automatic); date window at 3 o'clock
Estimated value: $3,640 ➔

1966
Memovox Polaris

Case: stainless steel, push-down case back, leather strap, Ø 42 mm
Movement: Caliber 425, rhodium-plated, 17 jewels, automatic winding
Remarks: rare alarm wristwatch with automatic winding (hammer automatic) for divers; water-resistant to 300 m; diving time set with rotating inner bezel (flange)
Estimated value: $15,600 ➔

ca. 1950
Deep Sea Alarm Automatic

Case: stainless steel, screw-down case back, leather strap, Ø 39 mm
Movement: Caliber 815, rhodium-plated, adjusted in 6 positions, 17 jewels, automatic with pendulum winding
Remarks: rare, water-resistant wrist alarm, signed "LeCoultre" (for the US market)
Estimated value: $10,400 ➔

1950
Chronograph

Case: stainless steel, screw-down case back, rotating bezel, leather strap, Ø 38 mm
Movement: Valjoux Caliber 726, gold-plated, column-wheel control of chronograph, manual winding
Remarks: sporty chronograph with 30-minute and 12-hour counters; only signed "Jaeger"
Estimated value: $2,340 ➔

1975
Club Automatic

Case: stainless steel, screw-down case back, leather strap, 38 x 43 mm
Movement: nickel-plated, automatic winding
Remarks: automatic men's watch with day and date display in the window at 3 o'clock
Estimated value: $1,040 ➔

1975
Memovox Automatic

Case: stainless steel, screw-down case back, stainless steel link bracelet, 38 x 45 mm
Movement: Caliber 916, rhodium-plated, automatic winding
Remarks: automatic alarm wristwatch with very solid case and stainless steel link bracelet
Estimated value: $1,300 ➔

1975
Memovox Automatic Snowdrop

Case: stainless steel, screw bezel, leather strap, Ø 43 mm
Movement: Caliber 916, rhodium-plated, automatic winding
Remarks: Snowdrop alarm wristwatch in design case; monocoque case closed via the bezel; date window at 3 o'clock
Estimated value: $1,950 ➔

JUNGHANS

More than one hundred years ago, this company was the largest watch manufacturer in the world. During its heyday, 12,000 watches poured from the workshops daily—and significant coordination and logistical effort were applied every day to pick up and bring more than 5,000 employees to the Black Forest's Schramberg to work.

Today, a workforce of less than one hundred is dwarfed by the giant factory building, which used to house one of Europe's most modern production parks. The company's history, the acme of which coincided with the advent of conveyor belt production and the later developed radio-controlled watch technology, mirrors the ups and downs of the German watch industry in its entirety.

The American production style was in high demand in the 1880s, allowing the clock company Junghans to finally compete with overseas suppliers of cheap clocks. The Black Forest–based company had manufactured cuckoo clocks since 1865, with Erhard Junghans progressively using the force of water for the efficient production of components made of brass, wood, and bronze. His brother Franz-Xaver had previously returned from the United States with modern machines and tools in tow.

The consistent introduction of "American" assembly line methods and the study of the superior watch technology of the large series manufacturers across the ocean helped Junghans to develop the company's first alarm clock, the famed Caliber 10. It became a milestone of cost-effective, simple design and was produced from 1885 to 1930.

An especially clever idea by Erhard's son Arthur in 1907 literally shed light on the subject. He mixed zinc sulfide with radium and applied the radioactively luminescent mixture to hands and dials so that they could be read the whole night through. What an innovation: one no longer needed to light a candle to see the time.

In 1903, Junghans launched its first pocket watch movement, and twenty-five years later its first wristwatch, thus entering a product segment that would continue to be extended in the ensuing years.

The company was partially dismantled by the Allies after World War II, but soon thereafter Junghans once again began developing and producing chronograph Caliber J88 and Minivox Caliber J89, which was patented in 1949.

Alongside chronographs and alarm wristwatches, Junghans manufactured simple watches outfitted with manually wound and automatic movements that were famous for the precision of their rates. In the 1950s, this company was the third-largest producer of chronometers after Rolex and Omega. After the Diehl concern of Nuremberg took over the majority of its stock in 1956, Junghans was divided into two departments: watches and precision engineering. During the quartz crisis of the 1970s, this proud name slipped into anonymity until the development of radio-controlled technology showed it the way out of its misery and secured the future of the brand. In 2000, Diehl sold Junghans's watch division to Egana-Goldpfeil Holding, headquartered in Hong Kong. Today it is once again privately owned and independent.

Founded:
1861 by Erhard Junghans in Schramberg

Headquarters:
Schramberg

Status:
active without interruption

Specialty:
diving watches, wristwatches, chronometers, pilot's chronographs, wristwatch alarms, dashboard watches and chronographs, radio-controlled watches, solar watches, kitchen clocks

Major inventions:
- 1907 introduction of radioactive luminescent mass
- radio-controlled wristwatches
- solar wristwatches

Leading models:
German Army Pilot's Chronograph, Minivox, Chronometer

1945
Chronograph

Case: chrome-plated, exhibition window case back, leather strap, Ø 36 mm
Movement: Caliber 88, gold-plated, frosted finish, column-wheel control of chronograph, manual winding
Remarks: chronograph with 30-minute counter and transparent case back
Estimated value: $1,040 ➔

1945
Chronograph

Case: stainless steel, screw-down case back, leather strap, Ø 38 mm
Movement: Caliber 88, gold-plated, frosted finish, column-wheel control of chronograph, manual winding
Remarks: chronograph with 30-minute counter
Estimated value: $1,300 ➔

1950
Chronograph

Case: chrome-plated, stainless steel push-down case back, leather strap, Ø 36 mm
Movement: Caliber 88, gold-plated, frosted finish, column-wheel control of chronograph, manual winding
Remarks: chronograph with 30-minute counter
Estimated value: $1,040 ➔

1950
Chronograph

Case: stainless steel, push-down case back, leather strap, Ø 36 mm
Movement: Caliber 88, gold-plated, frosted finish, column-wheel control of chronograph, manual winding
Remarks: chronograph with 30-minute counter
Estimated value: $1,300 ➔

1960
German Army Pilot's Chronograph

Case: chrome-plated, screw-down case back, leather strap, Ø 38 mm
Movement: Caliber 88, gold-plated, frosted finish, column-wheel control of chronograph, manual winding
Remarks: rare first edition of the German army pilot's chronograph featuring a fluted bezel; engraved with air force number 6645-12-120-9351/88-0110
Estimated value: $2,600 ➔

1960
German Army Pilot's Chronograph

Case: matte finish, stainless steel screw-down case back, rotating bezel, leather strap, Ø 38 mm
Movement: Caliber 88, gold-plated, frosted finish, column-wheel control of chronograph, manual winding
Remarks: pilot's chronograph of the German army, engraved with following inscription: "Bundeseigentum Nr.12-124-8591" ("property of the government" and number)
Estimated value: $2,340 ➔

1945
German Army Service Watch

Case: chrome-plated, stainless steel push-down case back, leather strap, Ø 36 mm
Movement: Caliber 80, nickel-plated, manual winding
Remarks: service wristwatch for the army's medical corps
Estimated value: $520 ➔

1940
Men's Watch

Case: stainless steel, push-down case back, 22 x 38 mm
Movement: Caliber 79/1, nickel-plated, manual winding
Remarks: rectangular men's watch
Estimated value: $260 ➔

1960
Minivox

Case: stainless steel, push-down case back, leather strap, Ø 35 mm
Movement: Caliber 89, nickel-plated, 20 jewels, manual winding
Remarks: alarm wristwatch with gong on case back; this watch was offered in its original box with certificate
Estimated value: $1,040 ➔

1960
Automatic with Power Reserve Indicator

Case: stainless steel, push-down case back, leather strap, Ø 35 mm
Movement: Caliber 80.12, nickel-plated, 22 jewels, automatic winding
Remarks: rare men's watch with power reserve indicator
Estimated value: $780 ➔

1955
Chronometer

Case: stainless steel, push-down case back, leather strap, Ø 33 mm
Movement: Caliber 82.1, gold-plated, manual winding
Remarks: chronometer
Estimated value: $780 ➔

1960
Chronometer Automatic

Case: gold-plated, stainless steel push-down case back, leather strap, Ø 34 mm
Movement: Caliber 83, gold-plated, 29 jewels, automatic winding
Remarks: chronometer
Estimated value: $1,040 ➔

1960
Chronometer Automatic

Case: 14-karat yellow gold, push-down case back, leather strap, Ø 36 mm
Movement: Caliber 83, gold-plated, automatic
Remarks: chronometer in gold case
Estimated value: $1,560 ➔

1960
Chronometer

Case: gold-plated, stainless steel push-down case back, leather strap, Ø 34 mm
Movement: Caliber 82.1, gold-plated, manual winding
Remarks: chronometer
Estimated value: $780 ➔

1965
Chronometer

Case: stainless steel push-down case back, leather strap, Ø 34 mm
Movement: Caliber 62.1, gold-plated, manual winding
Remarks: chronometer
Estimated value: $1,040 ➔

1965
Chronometer

Case: 14-karat yellow gold, push-down case back, leather strap, Ø 35 mm
Movement: Caliber 82.1, gold-plated, manual winding
Remarks: chronometer
Estimated value: $1,170 ➔

Audubon's Birds of America

The National Audubon Society
Baby Elephant Folio

"A brilliant achievement."
—*The New York Times*

Now with full-color illustrations throughout, this marvelous edition of *Audubon's Birds of America* displays all 435 of John James Audubon's brilliant hand-colored engravings in exquisite reproductions taken from the original plates of the Audubon Society's archival copy of the rare Double Elephant Folio. Organized and annotated by Roger Tory Peterson, America's best-known ornithologist, and issued with the full endorsement and cooperation of the Audubon Society, this magnificent volume is as thorough in scientific classification as it is beautiful. The colorful captions in the back provide fascinating commentaries on each featured bird.

By Roger Tory Peterson and Virginia Marie Peterson
482 full-color illustrations
694 pages · Cloth · $185.00
ISBN 978-0-7892-1135-4
E-BOOK ISBN 978-0-7892-6017-8

ABBEVILLE PRESS

Published by ABBEVILLE PRESS
137 Varick Street, New York, NY 10013
1-800-ARTBOOK (in U.S. only)
Also available wherever fine books are sold
Visit us at www.abbeville.com

LEMANIA

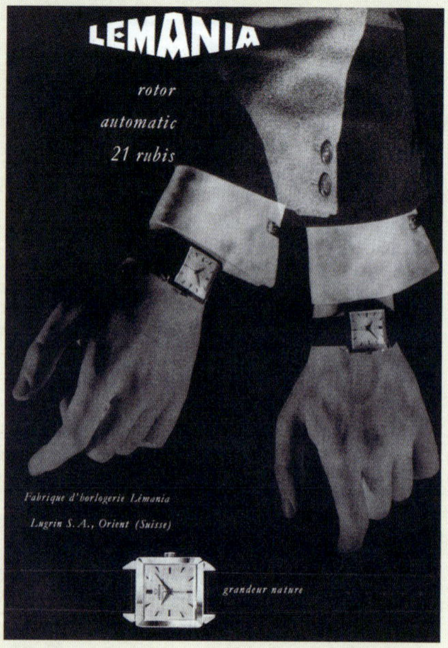

A certain Alfred Lugrin rented a small workshop in Le Sentier in 1883. There the self-taught watchmaker, who had worked his way up from assistant mechanic at Jaeger-LeCoultre to chronograph specialist, began to add chronograph mechanisms to purchased *ébauches* together with three colleagues. By 1885, Lugrin's company was already making complications like repeaters and chronographs; by 1890, it had grown so much that a generous factory had to be built in L'Orient, which today remains the heart of the oft-annexed modern facility. It was around 1900 that the production of complete watches under the company's own name began.

After Lugrin passed away, his son-in-law Marius Meylan took over managing the company. In 1930, an especially successful caliber inspired him to make its moniker that of the company: Lemania after Lac Léman, which is French for Lake Geneva. In 1932, just two years later, Lemania was integrated into the SSIH and became a chronograph supplier for the top brand Omega. Caliber 321 was exceptional; after all, it made it all the way to the moon inside the Speedmaster. In the 1970s, Lemania developed automatic chronograph calibers 1340, 1341, and 5100, which became true classics. The very robust Caliber 5100 remains for many the most reliable chronograph caliber ever created. The development of alarm wristwatch Caliber 980—a variation of which was exclusive to Omega and allowed setting the alarm to the minute—was precisely within this time frame.

The SSIH sold Lemania in 1980 to Piaget, and after a management buyout, the traditional company was relaunched as Nouvelle Lemania S.A. Then in 1992 the resuscitated watch brand Breguet, searching for a movement manufacturer, purchased and integrated it. It was from then on that the most complicated Breguet timepieces were created in the Nouvelle Lemania workshops. In the summer of 1999, the complete Breguet group, including Lemania, was purchased by the Swatch Group and restructured under the new name Breguet Manufacture. The Lemania name has now disappeared from the market.

Lemania stands for first-class pilot's chronographs in the hearts of wristwatch collectors, though the Lemania brand name also brought forth many interesting automatic men's watches in addition to the rare alarm wristwatches.

Founded:
1883 by Alfred Lugrin in Le Sentier

Headquarters:
L'Orient

Status:
merged into the Breguet Manufacture (Swatch Group)

Specialty:
development of chronograph calibers, chronographs, repeaters, tourbillons, pilot's watches

1920
Chronograph

Case: nickel, push-down case back, leather strap, Ø 32 mm
Movement: nickel-plated, côtes de Genève, column-wheel control of chronograph, manual winding
Remarks: early one-button chronograph with enamel dial
Estimated value: $1,560 ➔

1935
Chronograph

Case: stainless steel, push-down case back, leather strap, Ø 37 mm
Movement: gold-plated, frosted finish, column-wheel control of chronograph, manual winding
Remarks: chronograph with 30-minute counter
Estimated value: $1,300 ➔

1940
Chronograph

Case: stainless steel, screw-down case back, leather strap, Ø 35 mm
Movement: nickel-plated, column-wheel control of chronograph, manual winding
Remarks: rare chronograph with 30-minute and 12-hour counters
Estimated value: $1,430 ➔

1940
Chronograph

Case: stainless steel, tripartite, push-down case back, leather strap, Ø 37 mm
Movement: gold-plated, column-wheel control of chronograph, fine matte steel chronograph components
Remarks: sporty chronograph with 30-minute counter; telemeter scale on outer perimeter of dial; tachymeter scale in snall form in the middle of dial; navette-shaped chronograph buttons
Estimated value: $1,560 ➚

1935
German Air Force Pilot's Watch

Case: stainless steel, leather strap, 38 x 50 mm
Movement: nickel-plated, manual winding
Remarks: early pilot's watch for the German air force
Estimated value: $1,300 ➔

1938
German Air Force Pilot's Chronograph

Case: stainless steel, screw-down case back, leather strap, Ø 38 mm
Movement: gold-plated, frosted finish, column wheel control of chronograph, manual winding
Remarks: pilot's chronograph for the German air force signed RLM-NAV 5130 (Reich ministry of aerospace navigation)
Estimated value: $ 1,560 ➔

1935
Chronograph "Urania"

Case: stainless steel, push-down case back, leather strap, Ø 37 mm
Movement: rhodium-plated, polished steel chronograph parts, column-wheel control, 17 jewels, manual winding
Remarks: rare pilot's chronograph, made for Andreas Huber of Urania, Munich
Estimated value: $3,250 ➔

1950
Chronograph

Case: stainless steel, bipartite, screw-down case back, leather strap, Ø 35 mm
Movement: rhodium-plated, column-wheel control of chronograph, fine matte steel chronograph components
Remarks: sporty, elegant chronograph with 30-minute counter; two-tone dial with luminous numerals; minute scale underscored by a black ring; telemeter scale on outer perimeter
Estimated value: $1,300 ➔

1970
Royal Air Force Chronograph

Case: stainless steel, screw-down case back, leather strap, Ø 40 mm
Movement: Caliber 2220, nickel-plated, column wheel control of chronograph, manual winding
Remarks: pilot's chronograph for the Royal Air Force with 30-minute counter
Estimated value: $1,560 ➔

1965
Royal Air Force Chronograph

Case: stainless steel, bipartite, screw-down case back, leather strap, Ø 40 mm
Movement: Caliber LWC 2220, nickel-plated, column wheel control of chronograph, matte steel chronograph components, manual winding
Remarks: one-button pilot's chronograph for the British air force signed Royal Air Force 6BB/924-3306-2141/65, 30-minute counter. Outfitted with asymmetrical case to protect crown and buttons.
Estimated value: $1,560 ➔

1960
Swedish Air Force Pilot's Watch

Case: stainless steel, screw-down case back, leather strap, Ø 40 mm
Movement: Lemania Caliber LWO 2225, nickel-plated, manual winding
Remarks: pilot's watch for the Swedish air force (57/2461-TG195) with reset device for the second hand
Estimated value: $ 1,950 ➔

1955
Chronograph

Case: 18-karat rose gold, push-down case back, leather strap, Ø 36 mm
Movement: Caliber LWC 2520, rhodium-plated, finely polished steel chronograph parts, column-wheel control, glucydur screw balance with Incabloc shock protection, manual winding
Remarks: elegant chronograph with a 30-minute and 12-hour counter
Estimated value: $3,250 ➔

1948
Chronograph Monopoussoir

Case: stainless steel, screw-down case back, leather strap, Ø 38 mm
Movement: Caliber LWO 2210, gold-plated, frosted finish, finely polished steel chronograph parts, column-wheel control, compensation screw balance, 17 jewels, manual winding
Remarks: monopusher chronograph with 30-minute counter; made for the Hydrographic Institute of the Italian Navy in Genoa
Estimated value: $3,250 ➔

1975
Chronograph

Case: stainless steel, screw-down case back, leather strap, Ø 43 mm
Movement: Caliber 1873, nickel-plated, manual winding
Remarks: chronograph with 30-minute and 12-hour counters
Estimated value: $910 ➔

1971
Alarm Wristwatch

Case: gold-plated, bipartite, stainless steel screw-down case back, leather strap, 34 x 43 mm
Movement: Lemania Caliber LWO2980, rhodium-plated, automatic winding
Remarks: automatic alarm wristwatch with integrated gong; this movement has only one mainspring, which also provides the energy for the alarm
Estimated value: $780 ➔

1971
Alarm Wristwatch

Case: stainless steel, screw-down case back, leather strap, 44 x 38.6 mm
Movement: Lemania Caliber LWO 2980, nickel-plated, 19 jewels, automatic winding
Remarks: automatic alarm wristwatch in stainless steel case; alarm time can be set in either direction
Estimated value: $780 ➔

DODANE
1857
TYPE 23
OFFICIAL NATO SUPPLIER NO. F1950
The Dodane Type 23 stems from years of Air Force production where tolerance is zero. All pieces are chronographs and chronometer ready - many tested by the National Observatory of Measuring Time of Besançon. Dodane is a small manufacture run by the same family for over five generations.
TWI
TOTALLY · WORTH · IT
724 263 2286 · 201 894 4710
info@totallyworthit.com · www.TotallyWorthIt.com

LONGINES

Longines actually gets its name from the natural setting surrounding the factory: the building is located in Saint-Imier on fields that abut a little river called Les Longines.

This company was founded as Agassiz & Cie. in 1832 by Auguste Agassiz and his friends Henri Raiguel and Florian Morel. It began its career as an *établisseur,* assembling purchased components to make complete watches. Production was completed mainly by workers operating out of their homes; Agassiz & Cie. specialized in distribution.

In 1852, Agassiz passed the company's management on to his nephew, Ernest Francillon, who recognized that at this time it was especially robust, reasonably priced watches that were in demand. For Francillon, that primarily meant that he had to rationalize his watch production.

In 1866, he purchased the above-mentioned piece of property on the banks of the Schüss in Saint-Imier and promptly made the term *Les Longines* the brand name for his watches. Because the little river delivered free energy, Francillon was able to develop a machine-supported production process in the new factory, which consistently delivered reasonably priced movements and watches of high quality. Francillon was soon exporting his watches all over the world.

Longines became a synonym for professional pilot's watches thanks to the U.S. postal pilot Charles A. Lindbergh, who became the first man to fly across the Atlantic alone on May 21, 1927. Lindbergh had sketched his idea for a practical pilot's watch with navigational aid and sent it to John P.V. Heinmüller, then-director of Longines Wittnauer Watch Co. in the United States. Heinmüller was also president of the International Federation of Aviation and was an enthusiastic pilot himself, and thus he recognized the value of Lindbergh's idea. Heinmüller contacted Longines in Switzerland, where the Lindbergh Hour Angle Watch was brought to serial maturity. It only follows that the watches from Saint-Imier became the preferred equipment of pilots and adventurers. Aviation pioneers like Amelia Earhart, Howard Hughes, and Richard Evelyn Byrd put their trust in their Longines timekeepers.

Longines hit the high point of its fame in the 1950s, when the company even had a television show called *The Longines Chronoscope.* Its content was scientific topics, and new technical developments were explained and discussed with guests.

In 1912, Longines introduced an electromechnical system for starting and stopping chronographs. This system qualified the company to become the official timekeeper for big sports events and the Olympic Games. Good craftsmanship and high rate precision also characterized Longines wristwatches. The company's watchmakers often developed unusual solutions such as Caliber 340, whose rotor revolved on a toothed gear, keeping the movement very flat. Longines's automatic Caliber 980 with twin spring barrels was so technically advanced that it continues to be manufactured by Lemania today—in modified form and under another caliber number—for various Swatch Group brands. Longines chronographs were also technical gems: just look at Caliber 13 ZN, the first chronograph with flyback function—one that was a great step forward for pilots who didn't always have the time to push buttons for stopping and starting while they were navigating. Like many other brands, Longines lost its independence during the quartz crisis. At first, the brand joined ASUAG along with many other smaller brands, then via other mergers and buyouts, it became part of the Swatch Group.

Founded:
1832 by Ernest Francillon in St.-Imier

Headquarters:
St.-Imier

Status:
active without interruption (Swatch Group)

Specialty:
chronometers, railway watches, dashboard watches, chronographs, time-measuring systems

Major inventions:
- flyback switch for chronographs
- 1946 introduction of the vibrograph for regulating

Leading models:
Conquest, Admiral, Lindbergh, Weems

1925
Chronograph

Case: silver, silver cuvette, leather strap, Ø 35 mm
Movement: Caliber 13.33, nickel-plated, côtes de Genève, column-wheel control of chronograph, manual winding
Remarks: early chronograph with 30-minute counter; enamel dial
Estimated value: $3,900 →

1930
Doctor's Chronograph

Case: 14-karat yellow gold, gold cuvette, leather strap, Ø 35 mm
Movement: Caliber 13.33, gold-plated, frosted finish, column-wheel control of chronograph, manual winding
Remarks: early chronograph with 30-minute counter; enamel dial; central strap lugs
Estimated value: $6,500 →

1930
Chronograph

Case: yellow gold, leather strap, Ø 33 mm
Movement: nickel-plated, column-wheel control of chronograph, manual winding
Remarks: early chronograph with 30-minute counter; enamel dial
Estimated value: $7,150 →

1934
Chronograph

Case: 18-karat yellow gold, push-down case back, leather strap, Ø 34 mm
Movement: Caliber 13.33, gold-plated, frosted finish, column-wheel control of chronograph, manual winding
Remarks: chronograph with 30-minute counter; dial later replaced by Longines
Estimated value: $4,550 →

1936
Men's Watch

Case: 14-karat yellow gold, push-down case back, leather strap, Ø 32 mm
Movement: Caliber 27.0, gold-plated, frosted finish, jewels set in chatons, manual winding
Remarks: gold men's watch with enamel dial
Estimated value: $2,340 →

1934
Chronograph

Case: stainless steel, push-down case back, enamel dial, leather strap, Ø 37 mm
Movement: Caliber 13ZN, gold-plated, frosted finish, finely polished and beveled steel chronograph parts, column-wheel control, glucydur screw balance, manual winding
Remarks: extremely rare chronograph with 30-minute counter, made for the Hydrographic Institute of the Italian Navy
Estimated value: $7,800 ↗

1940
Chronograph

Case: 18-karat yellow gold, push-down case back, enamel dial, leather strap, Ø 37 mm
Movement: Caliber 30ZN, gold-plated, frosted finish, polished steel chronograph parts, column-wheel control, glucydur screw balance, manual winding
Remarks: extremely rare chronograph with 30-minute counter, made for the Hydrographic Institute of the Italian Navy
Estimated value: $6,500 ↗

1940
Lindbergh Hour Angle

Case: stainless steel, multipart, push-down case back with hinge, silver rotating bezel with hour angle function, enamel numeral ring, rotatable dial center, leather strap, Ø 46 mm
Movement: Caliber 37.9, rhodium-plated, jewels set in chatons, indirect sweep second, glucydur screw balance and anchor, manual winding
Remarks: large, rare large Lindbergh system pilot's navigation watch in stainless steel with a synchronizable second scale at the dial center and a graduated rotating silver bezel
Estimated value: $28,800 →

LONGINES CHRONOGRAPHS

THE UNRECOGNIZED MASTER

Nowadays, Longines is one of the most important brands within the Swatch Group. It basically meets just about every taste requirement in the medium price segment, from the lady's watch all the way to the pilot's chronograph. The models with a historic look from the Master, Legend, St.-Imier, and Conquest collections appear to be especially in demand. But for Longines connoisseurs, these watches all done up in a vintage look are nothing more than nicely packaged ETA mass production. There is no comparison with Longines watches from the days preceding the collectivization of the various Swiss brands under the roof of the ASUAG. The fact that Omega was built up to be the leading brand after merging into SMH (later Swatch Group) in 1983 was long a source of debate within the group, because Longines clearly had the higher-quality collection to show for itself and the more advanced technical development. But Omega was easier to pronounce.

There was a time when the brand name Longines stood for outstanding horological technology, top precision, and the finest manufacturing quality. Chronographs were the specialty of the brand with the winged hourglass on its logo—no wonder, since Longines was involved for years in measuring sports results at track-and-field competitions and the Olympic Games. The use of a wire in 1912 to ascertain the winner was a Longines invention.

And Longines was always in the forefront of chronograph technology, be it for the shift from the single pusher to double pusher mechanisms, the construction of flyback and stop seconds functions, and when optimizing column wheel control. Legendary chronograph calibers from the 1940s, like the 13 ZN or the 30 CH, are perfectly able to compete with the much-admired chronograph movements of the renowned Genevan *manufactures,* yet they are—at least for the moment—available at affordable prices.

1936
Men's Watch

Case: stainless steel, push-down case back, leather strap, 22 x 38 mm
Movement: Caliber 25.17, rhodium-plated, côtes de Genève, jewels set in chatons, manual winding
Remarks: early men's watch with sweep seconds; this watch was offered with the original Longines sales tag
Estimated value: $1,170 ➔

1935
Doctor's Watch

Case: stainless steel, push-down case back, leather strap, 20 x 30 mm
Movement: Caliber 9.32, rhodium-plated, côtes de Genève, jewels set in chatons, shaped movement, manual winding
Remarks: early men's watch in rectangular case; the large subsidiary seconds dial made pulse-taking easier
Estimated value: $1,300 ➔

1950
Hour Glass

Case: gold-plated, push-down case back, leather strap, 25 x 40 mm
Movement: Caliber 9 LT, nickel-plated, jewels set in chatons, manual winding
Remarks: gold men's watch in a fitted, shaped case reminiscent of an hourglass
Estimated value: $650 ➔

1954
Automatic

Case: 18-karat yellow gold, push-down case back, leather strap, 31 x 43 mm
Movement: Caliber 22A, nickel-plated, jewels set in chatons, automatic winding
Remarks: elegant men's watch with domed square case
Estimated value: $1,300 ➔

1942
Men's Watch

Case: 14-karat yellow gold, push-down case back, leather strap, 20 x 42 mm
Movement: Caliber 92, nickel-plated, jewels set chatons, manual winding
Remarks: men's watch with faceted crystal
Estimated value: $1,300 ➔

1947
Men's Watch

Case: 18-karat yellow gold, push-down steel case back, leather strap, 25 x 38 mm
Movement: Caliber 10L, rhodium-plated, jewels set in chatons, manual winding
Remarks: extravagant men's watch with accentuated strap lugs
Estimated value: $1,170 ➔

1939
Weems

Case: stainless steel, push-down case back, rotating bezel, leather strap, Ø 34 mm
Movement: Caliber 12.68N, gold-plated, frosted finish, manual winding
Remarks: early pilot's watch with rotating bezel according to the Weems principle
Estimated value: $1,950 ➔

1938
Weems

Case: stainless steel, push-down case back, rotating bezel, leather strap, Ø 33 mm
Movement: Caliber 12.68N, gold-plated, frosted finish, jewels set in chatons, manual winding
Remarks: rare navigational pilot's watch for the Royal Air Force; AM (Air Ministry) engraved on military case back
Estimated value: $1,950 ➔

1946
Chronograph

Case: stainless steel, Ø 37 mm, tripartite, push-down case back, leather strap
Movement: Caliber 1268Z, rhodium-plated, finely finished steel components
Remarks: one-button chronometer with central 60-minute counter; extra-large minute chapter ring printed on dial
Estimated value: $2,600 →

1940
Weems Navigation Watch

Case: stainless steel, multipart, push-down case back, enamel number ring, rotatable dial center, leather strap, Ø 48 mm
Movement: Caliber 37.9, rhodium-plated, indirect sweep second, glucydur screw balance and anchor, manual winding
Remarks: early and rare large pilot's navigation watch with the Weems system with a synchronizable second scale at the center of the dial
Estimated value: $15,600 →

1946
Weems Chronograph

Case: stainless steel, push-down case back, rotating bezel, leather strap, Ø 47 mm
Movement: rhodium-plated, côtes de Genève, manual winding
Remarks: pilot's watch with resettable sweep second and minute counters; rotating bezel with reference markings
Estimated value: $14,300 ↗

1936
German Air Force Pilot's Watch

Case: stainless steel, push-down case back, rotating bezel, leather strap, 40 x 51 mm
Movement: Caliber 15.94, gold-plated, frosted finish, manual winding
Remarks: pilot's watch for the young German air force
Estimated value: $1,950 →

1938
Lindbergh Hour Angle

Case: stainless steel, push-down case back, rotating bezel, leather strap, Ø 33 mm
Movement: Caliber 12L, rhodium-plated, jewels set in chatons, manual winding
Remarks: early, extremely rare Hour Angle navigational watch with rotating and locking bezel; this watch was offered with its original leather strap and silver buckle
Estimated value: $3,900 →

1950
Astro Navigation

Case: stainless steel, push-down case back, leather strap, Ø 47 mm
Movement: Caliber 37.9S, rhodium-plated, jewels set in chatons, manual winding
Remarks: extremely rare pilot's watch for Swissair with 24-hour display for astronavigation
Estimated value: $9,100 →

1934
Lindbergh Hour Angle

Case: silver, multipart, push-down case back with hinge, silver rotating bezel with hour angle function, enamel numeral ring, rotatable dial center, leather strap, Ø 47 mm
Movement: Caliber 18.69N, gold plating, frosted finish, indirect sweep second, compensation screw balance, glucydur anchor, manual winding
Remarks: early and rare large pilot's navigation watch with the Lindbergh system with a synchronizable second scale at the center of the dial and a graduated rotating bezel
Estimated value: $32,500 →

1935
Royal Serbian Air Force Pilot's Watch

Case: stainless steel, push-down case back, leather strap, Ø 37 mm
Movement: Caliber 15.26, gold-plated, frosted finish, manual winding
Remarks: rare pilot's watch for the Royal Serbian air force with military engraving on the case back
Estimated value: $2,600 →

1950
Royal Air Force

Case: stainless steel, bipartite, screw-down case back, textile strap, Ø 36 mm
Movement: Caliber 12.68Z, gold-plated, frosted finish, jewels set in chatons, blued screws, 15 jewels
Remarks: rare pilot's navigation watch for the Royal Army with subsidiary seconds; military engraving "W.W.W. F4780" on case back
Estimated value: 2,600 ➔

1951
Royal Air Force Pilot's Watch

Case: stainless steel, screw-down case back, leather strap, Ø 44 mm
Movement: Caliber 14.68N, gold-plated, frosted finish, jewels set in chatons, manual winding
Remarks: rare pilot's watch for the Royal Air Force
Estimated value: $3,250 ➔

1934
Chronograph

Case: stainless steel, two push-down cases with a hinge, leather strap, Ø 35 mm
Movement: Caliber 13ZN, gold-plated, frosted finish, polished steel chronograph parts, column-wheel control, glucydur screw balance, manual winding
Remarks: rare early pilot's watch with 30-minute counter
Estimated value: $5,850 ➚

1936
Chronograph

Case: stainless steel, Ø 36 mm, push-down case back, leather strap
Movement: Caliber 13ZN, rhodium-plated, fausses côtes decoration, column-wheel control of chronograph, fine matte chronograph steel components
Remarks: rare chronograph with 30-minute counter and minute chapter ring on dial perimeter; this watch was delivered to Longines, France
Estimated value: $7,800 ➔

1941
Military Chronograph

Case: stainless steel, screw-down case back, leather strap, Ø 37 mm
Movement: Caliber 13ZN, gold-plated, frosted finish, column-wheel control of chronograph, manual winding
Remarks: rare military pilot's chronograph with 30-minute counter
Estimated value: $11,700 ➔

1934
Chronograph

Case: stainless steel, push-down case back, leather strap, Ø 36 mm
Movement: Caliber 13ZN, gold-plated, frosted finish, polished steel chronograph parts, column-wheel control, glucydur screw balance, manual winding
Remarks: early chronograph with 30-minute counter and mobile lugs
Estimated value: $6,500 ➚

1934
Chronograph

Case: 18-karat yellow gold, push-down case back, leather strap, Ø 37 mm
Movement: Caliber 13ZN, gold-plated, frosted finish, polished steel chronograph parts, column-wheel control, glucydur screw balance, manual winding
Remarks: elegant chronograph with a 30-minute counter
Estimated value: $7,800 ➚

1934
Chronograph Monopoussoir

Case: stainless steel, engraved case back, leather strap, Ø 37 mm
Movement: Caliber 13ZN, gold-plated, finely polished and beveled steel chronograph parts, column-wheel control, compensation screw balance, manual winding
Remarks: pilot's watch with a monopusher and 30-minute counter
Estimated value: $5,200 ➚

1942
Chronograph

Case: stainless steel, Ø 36 mm, tripartite, push-down case back, leather strap
Movement: Caliber 13ZN, gold-plated, frosted finish, column-wheel control of chronograph, matte chronograph steel components
Remarks: chronograph with 30-minute counter; two-tone dial; blue tachymeter scale; red telemeter scale
Estimated value: $5,200 ➔

1934
Chronograph

Case: stainless steel, Ø 37 mm, tripartite, push-down case back, leather strap
Movement: Caliber 13ZN, gold-plated, frosted finish, column-wheel control of chronograph, matte steel components
Remarks: early chronograph with 30-minute counter; blue tachymeter and red telemeter scales
Estimated value: $5,200 ➔

1944
Chronograph

Case: stainless steel, push-down case back, leather strap, Ø 35 mm
Movement: Caliber 13ZN, rhodium-plated, côtes de Genève, column-wheel control of chronograph, manual winding
Remarks: heavy chronograph with 30-minute counter; this watch was offered in its original box and with papers
Estimated value: $10,400 ➚

1940
Men's Watch with Hacking Second/Flyback

Case: stainless steel, push-down case back, leather strap, Ø 37 mm
Movement: Caliber 12.68Z, rhodium-plated, jewels set in chatons, manually wound
Remarks: extremely rare men's watch with sweep seconds with flyback function and sweep minute counter
Estimated value: $2,600 ➔

1952
Chronograph

Case: 18-karat rose gold, Ø 37 mm, tripartite, push-down case back, leather strap
Movement: Caliber 30CH, rhodium-plated, fausses côtes decoration, jewels set in chatons, column-wheel control of chronograph, finest matte chronograph steel components
Remarks: rare chronograph with 30-minute counter and blue tachymeter scale; the case has especially striking strap lugs
Estimated value: $6,500 ➔

1956
"Doctor's Chronograph"

Case: stainless steel, push-down case back, leather strap, Ø 38 mm
Movement: Caliber 30CH, nickel-plated, jewels set in chatons, column wheel control, manually wound
Remarks: chronograph with extra pulsometer scale for quick pulse measurement
Estimated value: $7,800 ➔

1944
Chronograph

Case: 18-karat yellow gold, push-down case back, leather strap, Ø 38 mm
Movement: Caliber 13ZN, gold-plated, matte surface, column wheel control, manually wound
Remarks: rare chronograph
Estimated value: $5,850 ➔

1937
Chronograph

Case: 18-karat yellow gold, tripartite, push-down case back, leather strap
Movement: Caliber 13ZN, gold-plated, frosted finish, column-wheel control of chronograph, matte chronograph steel components
Remarks: rare simple chronograph with 30-minute counter
Estimated value: $4,850 ➔

1957
Chronograph

Case: 18-karat yellow gold, Ø 38 mm, bipartite, screw-down case back, gold link bracelet
Movement: Caliber 30CH, rhodium-plated, jewels set in chatons, column-wheel control of chronograph, finest matte steel components
Remarks: very rare chronograph with 30-minute counter and blue tachymeter scale; this watch was offered on a gold link bracelet
Estimated value: $7,800 ➔

1945
Chronograph

Case: stainless steel, Ø 39 mm, bipartite, screw-down case back, leather strap
Movement: Caliber 13ZN, rhodium-plated, fausses côtes decoration, column-wheel control of chronograph, fine matte chronograph steel components, 17 jewels
Remarks: extremely rare chronograph with central 60-minute counter and small 12-hour counter at 3 o'clock; chronograph minute hand highlighted in red
Estimated value: $ 26,000 ➔

1968
Military Chronograph

Reference number: 8226-2
Case: stainless steel, tripartite, screw-down case back, rotating bezel, leather strap, Ø 38 mm
Movement: Caliber 332, rhodium-plated, finely finished, column-wheel control of chronograph functions, finely finished steel chronograph components, 17 jewels
Remarks: pilot's chronograph with 30-minute and 12-hour counters; a second time zone can be set with the knurled rotating bezel
Estimated value: $4,160 ➚

1966
Chronograph

Case: stainless steel, Ø 36 mm, tripartite, push-down case back, leather strap
Movement: Caliber 30CH, rhodium-plated, jewels set in chatons, with column wheel control, finely matted steel chronograph parts, 18 jewels
Remarks: chronograph with 60-minute counter. The watch has an additional pulsation scale to facilitate pulse measurement.
Estimated value: $7,800 ➔

1957
Chronograph

Case: 18-karat red gold, Ø 37 mm, tripartite, push-down case back, leather strap
Movement: Caliber 30CH, rhodium-plated, jewels set in chatons, column-wheel control of chronograph, finest matte chronograph steel components, 18 jewels
Remarks: elegant chronograph with 30-minute counter and tachymeter scale; the strap lugs are very noticeable due to their shape
Estimated value: $5,850 ➔

1952
Chronograph

Case: 18-karat yellow gold, push-down case back, leather strap, Ø 35 mm
Movement: Caliber 30CH, nickel-plated, with côtes de Genève, manually wound
Remarks: gold chronograph with 30-minute counter
Estimated value: $6,500 ➔

1955
Antimagnetic Chronograph

Case: stainless steel, push-down case back, leather strap, Ø 37 mm
Movement: Caliber 30CH, nickel-plated, ribbing, column wheel control, manually wound
Remarks: chronograph with 30-minute counter and reworked dial
Estimated value: $5,200 ➔

1951
Chronograph

Case: stainless steel, Ø 38 mm, bipartite, screw-down case back, leather strap
Movement: Caliber 30CH, rhodium-plated, jewels set in chatons, column wheel control, finely matted steel chronograph parts, 18 jewels
Remarks: chronograph with 30-minute counter, blue tachymeter and red telemeter scales
Estimated value: $5,850 ➔

1953
Chronograph

Reference number: 6592-2
Case: stainless steel, screw-down case back, leather strap, Ø 38 mm
Movement: Caliber 30CH, rhodium-plated, decoration, polished steel chronograph parts, column-wheel control, glucydur screw balance, manual winding
Remarks: chronograph with 30-minute counter
Estimated value: $4,550 ↗

1956
Chronograph

Reference number: 7414/10
Case: 18-karat yellow gold, push-down case back, leather strap, Ø 36 mm
Movement: Caliber 530, rhodium-plated, frosted finish, finely polished and beveled steel chronograph parts, column-wheel control, glucydur screw balance with shock protection, manual winding
Remarks: very fine, rare chronograph with a 30-minute counter as well as a tachymeter and telemeter scale
Estimated value: $5,850 ↗

1960
Chronograph

Case: stainless steel, push-down case back, leather strap, Ø 34 mm
Movement: Caliber 332, nickel-plated, column wheel control, 17 jewels, manually wound
Remarks: chronograph with 30-minute and 12-hour counter
Estimated value: $1,820 →

1965
Automatic Diver's Watch

Case: stainless steel, screw-down case back, leather strap, Ø 42 mm
Movement: Caliber 290, nickel-plated, automatic
Remarks: automatic diver's watch with a rotating bezel inside (flange)
Estimated value: $5,200 →

1972
Conquest "XXth Olympic Games 1972"

Case: stainless steel, screw-down case back, leather strap, Ø 36 mm
Movement: Caliber 334, nickel-plated, column wheel control, manually wound
Remarks: single-pusher chronograph, special model for the 1972 Olympic Games in Munich; Olympia engraving on the case back
Estimated value: $780 →

1975
Chronograph

Case: stainless steel, screw-down case back, leather strap, 44 x 44 mm
Movement: Caliber Valjoux 726, nickel-plated, column wheel control, manually wound
Remarks: heavy chronograph with 30-minute and 12-hour counters
Estimated value: $1,820 →

1975
Chronograph with Nonius Hand

Reference number: 8723-1
Case: 18-karat yellow gold, screw-down case back, leather strap, 38 x 42 mm
Movement: Caliber Valjoux 72, rhodium-plated, column-wheel control, glucydur screw balance with shock protection, manual winding
Remarks: rare gold chronograph with 12-hour and 30-minute counter as well as a second hand with Nonius scale to read 1/10 seconds
Estimated value: $7,800 ↗

1978
UltraChron

Case: stainless steel, screw-down case back, stainless steel link bracelet, Ø 43 mm
Movement: Caliber 431, nickel-plated, automatic
Remarks: diver's watch with inner rotatable bezel (flange)
Estimated value: $1,170 →

1978
Comet

Case: stainless steel, screwed-down case back, stainless steel link bracelet, 34 x 41 mm
Movement: Caliber 702, rhodium-plated, manually wound
Remarks: rare men's watch with time displayed on rotating disks
Estimated value: $1,300 →

1975
Conquest Automatic

Case: stainless steel, screw-down case back, leather strap, Ø 43 mm
Movement: Caliber 431, nickel-plated, automatic
Remarks: diver's watch with inner rotatable bezel (flange), 36,000 vph high-frequency movement
Estimated value: $1,170 →

1950
Mystérieuse

Reference number: 167-B
Case: 18-karat white gold, diamond bezels and lugs, push-down case back, leather strap, Ø 32 mm
Movement: Caliber 19.4, rhodium-plated, glucydur screw balance, 17 jewels, manual winding
Remarks: hour markers and index on the hour disc in the dial center set with diamonds
Estimated value: $2,340 ↘

1960
Conquest Automatic Power Reserve

Reference number: 9032
Case: stainless steel, screw-down case back, leather strap, Ø 35 mm
Movement: Caliber 294, nickel-plated, 24 jewels, automatic winding
Remarks: extremely rare automatic jewel bearing, rotor-driven timepiece; central 45-hour power reserve indicator; date
Estimated value: $1,560 →

1960
Conquest Calendar Automatic

Case: 18-karat yellow gold, screw-down case back, leather strap, Ø 35 mm
Movement: Caliber 19ASD, rhodium-plated, jewels set in chatons, 19 jewels, automatic winding
Remarks: automatic men's watch with date display; this watch was offered in its original box
Estimated value: $1,950 →

1960
Conquest Automatic

Case: 18-karat yellow gold, screw-down case back, leather strap, Ø 35 mm
Movement: Caliber 19AS, nickel-plated, jewels set in chatons, 19 jewels, automatic winding
Remarks: heavy gold men's watch in original box
Estimated value: $1,820 →

1965
Flagship Automatic

Reference number: 3503
Case: yellow gold, push-down case back, leather strap, Ø 35 mm
Movement: Caliber 340, gold-plated, 17 jewels, automatic winding
Remarks: automatic men's watch with unusual winding mechanism utilizing a toothed rim underneath the rotor
Estimated value: $1,560 →

1965
Grand Prize Automatic

Case: 14-karat yellow gold, screw-down case back, leather strap, Ø 34 mm
Movement: Caliber 340, rhodium-plated, automatic winding
Remarks: automatic men's watch with applied, faceted gold markers; unusual winding mechanism utilizing a toothed rim underneath the rotor
Estimated value: $1,560 →

MINERVA

The owners of this company chose to name their enterprise for Minerva, the Roman goddess of the arts and crafts. They envisioned the goddess holding her hands over their small business, one that would soon create a good reputation for itself with excellent chronograph movements. The company was actually founded in 1858 in Villeret by the brothers Charles-Yvan and Hippolyte Robert as a small factory for assembling movements long before these owners asked the Roman deity for her aid. The *ébauches* were purchased from the factory known as FHF in Fontainemelon. Twenty years later, Charles and Georges Robert took over the management of this *établisseur* factory, and in 1885 their brother Yvan-Robert joined them as the third manager. The three chose an arrow-shaped version of the letters RF for Robert Frères as their logo and had it registered in 1887.

Around the turn of the century, they were experiencing so much success that they even ventured to manufacture their own calibers for pocket watches. In 1908, the Robert brothers presented their first movement to include a chronograph function, Caliber 9CH. With this, they laid the cornerstone for their future reputation as manufacturers of high-quality chronograph calibers.

The year 1923 was one of the most important in the company's history for a number of reasons. First, the company's owners changed its name to Fabrique Minerva, Robert Frères SA, and second, they presented their first chronograph caliber for wristwatches, Caliber 13-20 CH. The watches that they outfitted with this new movement became the leading models for the small *manufacture,* but they also supplied other companies.

It might have appeared that the goddess Minerva had turned her back on the company in the 1930s, as it was subjected to the full brunt of the world economic crisis, then in full swing. The founding family left the company and sold it to watch technician Jacques Pelot and mechanic Charles Haussener. The pair fought with all their might for the brand's survival and developed numerous new calibers. Under their management, Minerva specialized more and more in chronographs. In 1940, Pelot's nephew, André Frey, joined the company and soon bought the whole thing. Under his management, one of the most important Minerva movements, Caliber 48 with subsidiary seconds, was developed. Frey passed away in September 2004, four years after the sale of the company to an Italian investment group planning a great comeback.

In late 2006, Richemont took it over, and today Minerva makes movements exclusively for the group's Montblanc brand.

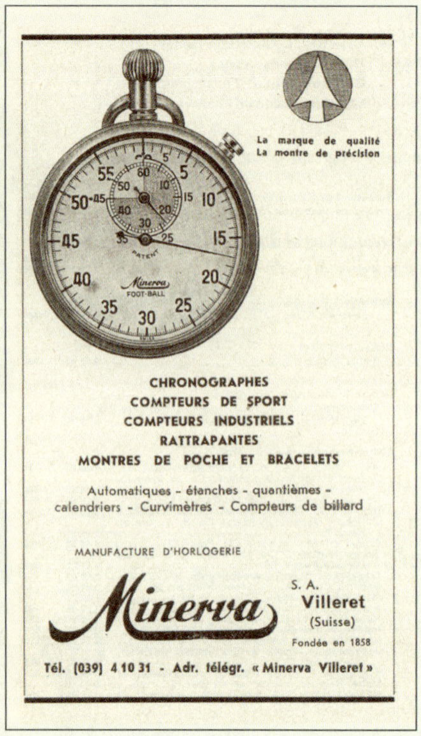

Founded:
1858 by the Robert brothers in Villeret

Headquarters:
Villeret

Status:
active without interruption (today Montblanc SA/Richemont Luxury Group)

Specialty:
chronographs

Leading models:
Pythagore

1935

Chronograph

Case: stainless steel, push-down case back, leather strap, Ø 37 mm
Movement: gold-plated, frosted finish, finely polished steel chronograph parts, column-wheel control, screw balance, manual winding
Remarks: chronograph with 12-hour and 30-minute counter
Estimated value: $2,600 ➔

1930

Chronograph Monopoussoir

Case: stainless steel, gold-plated, push-down case back with hinge, leather strap, Ø 45 mm
Movement: Caliber Minerva, rhodium-plated, fausses côtes decoration, finely matted steel chronograph parts, column-wheel control, screw balance, manual winding
Remarks: pilot's watch with a monopusher chronograph and 30-minute counter
Estimated value: $4,550 ➔

1935

Chronograph

Case: stainless steel, Ø 42 mm, tripartite, push-down case back, leather strap
Movement: rhodium-plated, fausses côtes decoration, column-wheel control of chronograph, fine matte chronograph steel components, manual winding
Remarks: early large one-button chronograph with 30-minute counter and tachymeter scale
Estimated value: $3,640 ➔

1938

German Air Force Chronograph

Case: stainless steel, screw-down case back, leather strap
Movement: gold-plated, frosted finish, manual winding, column-wheel control of chronograph
Remarks: rare pilot's chronograph for the German air force with 30-minute counter and tachymeter scale
Estimated value: $3,900 ➔

1943

Split-Seconds Chronograph

Case: stainless steel, push-down case back, leather strap, Ø 43 mm
Movement: nickel-plated, côtes de Genève, double column-wheel control of chronograph, manual winding
Remarks: rare pilot's chronograph with split seconds function, 30-minute counter; luminous numerals
Estimated value: $15,600 ➔

1940

Chronograph

Case: 18-karat yellow gold, Ø 34 mm, tripartite, push-down case back, leather strap
Movement: rhodium-plated, fine matte steel chronograph components, polished screws
Remarks: elegant gold chronograph with 30-minute counter
Estimated value: $3,510 ➔

1950

Chronograph

Case: stainless steel, Ø 36 mm, bipartite, screw-down case back, leather strap
Movement: rhodium-plated, fausses côtes decoration, column-wheel control of chronograph, fine matte steel chronograph components
Remarks: sporty chronograph with 30-minute and 12-hour counters; complete calendar with date hand; window display for weekdays and months
Estimated value: $4,550 ➔

1965

Chronograph

Case: stainless steel, Ø 36 mm, tripartite, screw-down case back, leather strap
Movement: Valjoux Caliber 72, rhodium-plated, decorated, column-wheel control of chronograph, fine matte steel chronograph components
Remarks: sporty chronograph with 30-minute and 12-hour counters
Estimated value: $3,900 ➔

MOVADO

Movado proved its competence in movements in momentous fashion in 1969 when it joined Zenith in introducing the world's first automatic chronograph.

Movado had been founded in 1881 in Le Chaux-de-Fonds by Alsatian Achille Ditisheim, who began with the manufacture of pocket watches in his little workshop. Convinced of their brother's quick success, Léopold and Isidor soon joined him. The three of them founded L.A.I. Ditisheim. For the purpose of recognition, they had registered different names to sell their watches under. The most successful of these came to the fore in 1903: Movado. This word was taken from the artificial world language Esperanto and means "always in motion." The easy pronunciation in various languages lent it high recognition. When in 1905 the fourth brother, Isaac, joined the company, it was renamed Ditisheim & Frères.

Seemingly inspired by the name Movado, the company also tried to go down new paths with its movements and watches. Attention to detail was visible especially on the movements. All bridges and cocks were designed in either strict geometrical or flowingly harmonious styles. In 1912, Movado presented the Polyplan, the shaped movement of which had a base plate that was rounded on both sides. The mechanism's curved shape ensured that it perfectly fit the strongly domed watch case. In 1914, Movado presented a trench watch that featured an additional grill over the sensitive crystal. The grill could be raised by a button located on the case at 6 o'clock so that the soldier could more easily read the time.

Before attacking the development of its own true chronograph, Movado introduced its Chronoplan in 1937, a short-term timer precise to the minute with a double rotating bezel. Outfitted with two reference markers, one turned to the position of the hour hand and the other to that of the minute hand, longer measurements of time could easily be read using the watch's main hands.

In developing its own chronograph caliber, Movado also went its own way. Instead of making a completely new movement, the developers took a tried-and-true manually wound caliber and built a chronograph module for it, which was positioned on the back of the slightly altered mechanism. Caliber 90M's main innovation lay in its counter: instead of the usual 30-minute totalizer, the movement was outfitted with a 60-minute counter at 3 o'clock. The new chronograph, presented in 1939, gave birth to an evolution, Caliber 95M, which featured an additional 12-hour counter.

Without a doubt the Museum Watch is the best-known of all Movado's models: it owes its name to its permanent spot in the Museum of Modern Art in New York. This simple watch, the dial of which is only decorated by a so-called sun dot at 12 o'clock, was created by American designer Nathan George Horwitt in 1947. Collectors also still search out the ergonomically domed Polyplan watches with shaped movements and a crown at 12 o'clock.

In 1969, Movado lost its entrepreneurial independence and merged with Zenith. In 1984, the brand was sold to the North American Watch Corporation, now MGI Luxury Group, a company that also owns the brands Concord and Ebel, where it remains to this day.

Founded:
1881 by Achille Ditisheim in La Chaux-de-Fonds

Headquarters:
La Chaux-de-Fonds

Status:
active without interruption (today MGI Luxury Group)

Specialty:
chronometers, diver's watches, railway watches, design watches

Major inventions:
- 1912 Polyplan
- 1931 Curviplan
- 1960 "Museum Watch" with a Nathan G. Horwitt dial
- 1969 first automatic chronograph "El Primero"

Leading models:
Chronoplan, Polyplan, Curvex, Astronic, Datron

1920
Chronomètre

Case: Sterling silver, push-down case back, leather strap, 26 x 45 mm
Movement: rhodium-plated, côtes des Genève, jewels set in chatons, 15 jewels, manual winding
Remarks: very early wristwatch (approx. 1920) in navette-shaped case; two soldered silver wires serve as the strap lugs
Estimated value: $1,300 →

1940
Polyplan

Case: 18-karat white gold, domed, push-down case back with hinge, leather strap, 22 x 54 mm
Movement: Caliber 400, rhodium-plated, polished, screw balance, 15 jewels, manual winding
Remarks: movement has a doubly angled mainplate and drive gears with inclined teeth to fit inside the curved case
Estimated value: $6,500 ↘

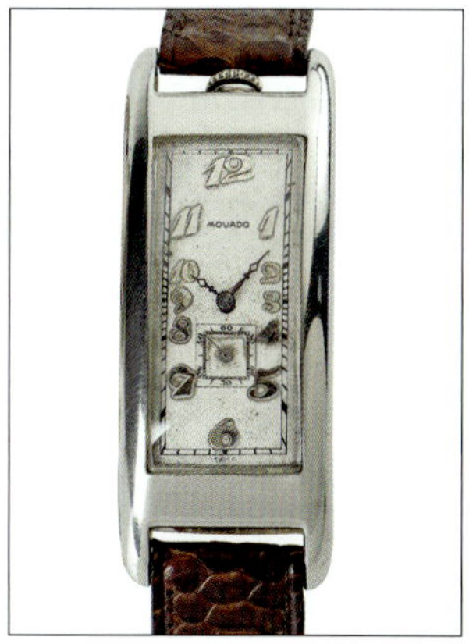

1935
Pilot's Watch Chronomètre

Case: stainless steel, push-down case back, rotating bezel, Ø 38 mm
Movement: Movado Caliber 75, rhodium-plated, jewels set in chatons, 15 jewels, manual winding
Remarks: extremely rare early pilot's watch; certified chronometer; movement adjusted in four positions
Estimated value: $3,250 →

1945
Tempomatic

Case: stainless steel, push-down case back, leather strap, Ø 37 mm
Movement: Movado Caliber 226, nickel-plated, automatic winding
Remarks: men's watch with hammer automatic; this watch was offered in its original box
Estimated value: $1,560 →

1945
Curvex

Case: stainless steel, leather strap, 23 x 41 mm
Movement: nickel-plated, côtes de Genève, jewels set in gold chatons, 15 jewels, manual winding
Remarks: rare men's watch in a domed Curvex case; watch with double signature Huber/Movado, which leads to the assumption that this watch is a special edition for Munich retailer Huber; movement adjusted in four positions
Estimated value: $1,300 →

1950
Calendoscope

Case: gold-plated, stainless steel push-down case back, leather strap, Ø 34 mm
Movement: Movado Caliber 123, rhodium-plated, 15 jewels, manual winding
Remarks: simple men's watch with date window at 3 o'clock
Estimated value: $260 →

1950
Chronograph

Case: stainless steel, push-down case back, leather strap, Ø 36 mm
Movement: Movado Caliber C95M, nickel-plated, with column-wheel control of chronograph, manual winding
Remarks: simple chronograph with 60-minute and 12-hour counters; tachymeter scale
Estimated value: $2,600 →

1945
Chronograph

Case: stainless steel, push-down case back, leather strap, Ø 35 mm
Movement: Movado Caliber 90, rhodium-plated, column-wheel control of chronograph, manual winding
Remarks: simple wristwatch chronograph with off-center 60-minute counter and tachymeter scale
Estimated value: $3,250 →

1945
Chronograph

Case: stainless steel, screw-down case back, leather strap, Ø 33 mm
Movement: Movado Caliber 90, nickel-plated, decorated, column-wheel control of chronograph, manual winding
Remarks: men's chronograph with off-center 60-minute counter, unusual for a chronograph
Estimated value: $3,250 ➔

1948
Doctor's Watch

Case: 14-karat yellow gold, screw-down case back, leather strap, Ø 33 mm
Movement: Movado Caliber 470, nickel-plated, manual winding
Remarks: rare so-called Doctor's Chronograph: simple hacking mechanism with reset possibility; pulsometer for easier pulse-taking
Estimated value: $3,900 ➔

1950
Calendar Watch

Case: 18-karat red gold, push-down case back, leather strap, Ø 34 mm
Movement: nickel-plated, jewels set in chatons, manual winding
Remarks: simple red gold watch with subsidiary seconds and complete calendar; sweep date hand; window displays for weekday (9 o'clock) and month (3 o'clock)
Estimated value: $2,600 ➔

1945
Calendomatic

Reference number: 6370
Case: 18-karat yellow gold, screw-down case back, leather strap, Ø 35 mm
Movement: Caliber 233, rhodium-plated, polished, 17 jewels, glucydur screw balance with shock protection, automatic winding with an oscillating weight
Remarks: full calendar with automatic winding
Estimated value: $3,250 ➔

1950
Chronograph

Reference number: 19038
Case: stainless steel, screw-down engraved case back, leather strap, Ø 35 mm
Movement: Caliber 190, rhodium-plated, finely polished and beveled steel chronograph parts, column-wheel control, glucydur screw balance, manual winding
Remarks: chronograph with 12-hour and 60-minute counter
Estimated value: $3,250 ➔

1955
Automatic

Case: stainless steel, screw-down case back, leather strap, Ø 33 mm
Movement: Movado Caliber 202, nickel-plated, automatic winding
Remarks: simple steel watch with exceptional date display; both the date and the weekday are displayed in the subsidiary seconds dial at 6 o'clock; hammer automatic movement
Estimated value: $1,300 ➔

1950
Calendar Watch with Moon Phase

Case: 18-karat rose gold, screw-down case back, leather strap, Ø 34 mm
Movement: Movado Caliber 473, nickel-plated, manual winding
Remarks: men's watch with subsidiary seconds, complete calendar and moon phase; sweep date hand; window displays for weekday and month; moon phase display located in subsidiary seconds dial
Estimated value: $ 3,250 ➔

1950
World Time

Case: stainless steel, push-down case back, rotating bezel with reference cities of the 24 world time zones, leather strap, Ø 36 mm
Movement: rhodium-plated, polished, 15 jewels, glucydur screw balance with shock protection, manual winding
Remarks: additional freely adjustable 24-hour hands to display the second time zone
Estimated value: $3,900 ➔

1960
Chronograph

Case: stainless steel, screw-down case back, leather strap, Ø 34 mm
Movement: Movado Caliber 95, nickel-plated, decorated, column-wheel control of chronograph, manual winding
Remarks: wristwatch chronograph with off-center 60-minute and 12-hour counters
Estimated value: $3,250 ➔

1965
Chronograph Super Sub Sea

Reference number: 206-704-501
Case: stainless steel, screw-down case back, rotating bezel with 60-minute division, leather strap, Ø 41 mm
Movement: Caliber 146HP, rhodium-plated, finely polished steel chronograph parts, column-wheel control, glucydur screw balance with shock protection, manual winding
Remarks: water-resistant chronograph with 12-hour and 30-minute counter
Estimated value: $2,600 ➔

1975
Chronograph Datron HS 360

Case: stainless steel, screw-down case back, leather strap, Ø 37 mm
Movement: Movado Caliber 3019PHC, nickel-plated, column-wheel control of chronograph, automatic winding
Remarks: heavy automatic chronograph outfitted with the legendary Zenith El Primero beating at 36,000 vph; unusual arrangement of date window at 12 o'clock
Estimated value: $1,300 ➔

1975
Super Sub Sea Datron HS360

Case: stainless steel, screw-down case back, rotating bezel, leather strap, Ø 42 mm
Movement: Movado Caliber 3019PHC, nickel-plated, column-wheel control of chronograph, automatic winding
Remarks: heavy automatic chronograph outfitted with the legendary Zenith El Primero beating at 36,000 vph; unusual arrangement of date window at 12 o'clock
Estimated value: $2,600 ➔

1960
Automatic

Case: 18-karat yellow gold, push-down case back, leather strap, Ø 35 mm
Movement: Movado Caliber 115, silver-plated, automatic winding
Remarks: simple men's watch with hammer automatic and subsidiary seconds; unusual dial with guilloché zone in the center
Estimated value: $1,300 ➔

1965
Men's Watch

Case: gold-plated, push-down case back, leather strap, Ø 34 mm
Movement: Movado Caliber 365, nickel-plated, jewels set in chatons, manual winding
Remarks: remarkably designed dial
Estimated value: $390 ➔

1975
Chronograph Datron HS360

Case: stainless steel, screw-down case back, stainless steel link bracelet, Ø 43 mm
Movement: Movado Caliber 3019PHC, nickel-plated, column-wheel control of chronograph, automatic winding
Remarks: heavy automatic chronograph outfitted with the legendary Zenith El Primero beating at 36,000 vph; unusual arrangement of date window at 12 o'clock
Estimated value: $1,820 ➔

1975
Chronograph Astronic

Case: stainless steel, screw-down case back, stainless steel link bracelet, 38 x 45 mm
Movement: Movado Caliber 3019PHF, rhodium-plated, column-wheel control of chronograph, automatic winding
Remarks: chronograph (Caliber El Primero) with 30-minute and 12-hour counters; complete calendar and moon phase
Estimated value: $2,600 ➔

OMEGA

Today it may be hard to believe, but watch giant Omega began as a one-man operation. In 1848, twenty-three-year-old Louis Brandt founded a so-called *comptoir d'établissage* in La Chaux-de-Fonds, a workshop for assembling watch components that included a distribution and sales department. The success of Brandt's watches was not long in coming thanks to their attractive appearance and high quality. These timekeepers were in demand in foreign countries, and Brandt spent weeks traveling through Europe to ensure delivery. Brandt's sons Louis Paul and César continued the active distribution policy of their father and moved the company headquarters to Biel in 1880. One of the reasons for the transfer was the city's better travel options. Another lay in the fact that Biel's canals offered waterpower free of charge, something that was of great importance to a mechanical workshop at the time. Finally, the Brandt brothers planned to manufacture their own watches outfitted with their own movements. They successfully realized their goals, and by 1889 their company was the largest watch factory in Switzerland.

Although the Brandts joined forces with other manufacturers now and then for individual models, the company's focus remained the development of autonomous watches and movements. In 1894, an especially successful caliber, which was far ahead of most of its competitor products thanks to its simple design and the exchangeability of all components, lent the company the name it would carry to the present day: Omega, the last letter of the Greek alphabet. All other brand names used by the company up to that point had been dropped by 1903, and Paul-Emile Brandt took over the management of Omega, retaining control for fifty years.

Paul-Emile Brandt's top priority was always precision—for which reason in 1917 the British air force took an Omega on as its official service watch, with the American army following suit one year later. In 1919, an Omega won the precision competition of the Neuchâtel Observatory for the first—but not the last—time. This success became the basis for a new watch line, the Constellation Chronometer, and for many years this model remained at the summit of the chronometer tests. In 1967, Omega received its one-millionth official rate certificate for a chronometer. The numerous models containing the legendary 30-millimeter caliber are a chapter in watch history in and of themselves and were sold for decades in various incarnations. They dominated the official chronometer competitions at observatories for a long time.

The most famous of all Omega models is surely the Speedmaster. This somewhat understated chronograph actually became official equipment for NASA astronauts thanks to its robustness and precision and was thus already a legend before taking part in the first moon landing. This brand even made a name for itself in the area of sports timing as well. Omega was and is still the official timekeeper for important sporting events such as the Olympic Games.

Omega has covered just about every niche with its watches. An automatic alarm wristwatch that could set the wake-up time precisely to the minute was even in the program: Memomatic. The Biel-based company did not shy away from the world of water, either, and created a diver's watch line with the tell-all name Seamaster. The most professional of the Omega diver's watches is model 600, also called Ploprof (for *plongeur professionel,* or "pro diver"). This watch proved itself on the wrists of professional divers employed by Comex in 252-meter depths. Even more popular is the Seamaster Professional, which can even boast its own film and television career. In the 1920s, Omega merged with Tissot to become SSIH (Société Suisse de l'Industrie Horlogère). Combining this with ASUAG (which also owned Longines), under the leadership of Nicolas G. Hayek, the SMH (Société Suisse de Microélectronique et Horlogerie) was created, which later became the Swatch Group.

Founded:
1848 by Louis Brandt in La Chaux-de-Fonds

Headquarters:
Biel/Bienne

Status:
active without interruption (Swatch Group)

Specialty:
chronometers, chronographs, diver's watches, railway watches, pilot's watches, wristwatch alarms, jewelry watches

Major inventions:
among others,
- 1910 time measurement with a trigger string
- 1967 Beta 21 tuning fork
- 1999 serial production of the coaxial escapement

Leading models:
Seamaster, Speedmaster, Constellation, De Ville

1935
Marine

Case: stainless steel, push-down case back, leather strap, 24 x 34 mm
Movement: Caliber 19.4 T2, silver-plated, manual winding
Remarks: rare early water-resistant men's watch with double case
Estimated value: $3,900 ➔

1935
Marine

Case: 14-karat yellow gold, 36 x 23 mm, tripartite, leather strap
Movement: rhodium-plated, finely finished, polished screws, 15 jewels
Remarks: rare wristwatch in Omega's patented marine case; this watch was produced under the signature "Udall & Ballou" for the jeweler of the same name in New York
Estimated value: $5,200 ➔

1934
Men's Watch

Case: stainless steel, push-down steel case back, leather strap, 25 x 39 mm
Movement: Caliber T17, nickel-plated, shaped movement, manual winding
Remarks: early Omega men's watch
Estimated value: $1,040 ➔

1946
Cosmic

Reference number: 2471/1
Case: stainless steel, tripartite, push-down case back, leather strap, Ø 35 mm
Movement: Caliber 27DLPC, red gold-plated, polished screws, 17 jewels
Remarks: steel calendar watch with complete calendar, date hand, window displays of weekday and month; subsidiary seconds and moon phase display at 6 o'clock
Estimated value: $4,550 ➔

1946
Cosmic

Reference number: 2473
Case: 14-karat yellow gold, tripartite, push-down case back, Ø 34 mm
Movement: Caliber 27DLPC, red gold-plated, polished screws, 17 jewels
Remarks: rare beautiful calendar watch with complete calendar, date hand, window displays of weekday and month; subsidiary seconds and moon phase display at 6 o'clock
Estimated value: $4,500 ➔

1936
Men's Watch

Case: 18-karat red gold, push-down case back, leather strap, Ø 37 mm
Movement: Caliber 30 T2 PC, red gold-plated, manual winding
Remarks: elegant men's watch
Estimated value: $3,250 ➔

1946
Chronomètre

Case: 18-karat yellow gold, tripartite, push-down case back, leather strap, Ø 33 mm
Movement: Caliber 30 T2 RG, red gold-plated, 17 jewels, regulated in 6 positions
Remarks: gold chronometer with Ø 30 mm caliber
Estimated value: $5,200 ➔

1947
Chronometer

Reference number: 30 T2 SCRG
Case: 18-karat yellow gold, tripartite, push-down case back, leather strap, Ø 35 mm
Movement: Caliber 30 T2 SCRG, red gold-plated, mirror-polished screws, gold screw compensation balance, fine adjustment, 16 jewels
Remarks: elegant chronometer with Ø 30 mm caliber
Estimated value: $5,200 ➔

OMEGA SPEEDMASTER PROFESSIONAL

THE LEGEND LIVES

The space adventure of the Omega Speedmaster began in 1964 with an anonymous shopping tour by two NASA employees to the jeweler named Corrigan in Houston. They purchased five chronographs from various brands with the aim of testing them. A short time later, NASA ordered twelve Speedmasters for $82.50 each from the Omega importer in New York. The watch had passed the test. In March 1965, the two first certified Speedmasters took off for orbit around the earth on the wrists of the astronauts Virgil Grissom and John Young. The model's first space walk took place in June on the arm of Edward White. He wore the watch on a long textile band fastened over his space suit with Velcro.

Omega only got wind of this over half a year later, in April 1966 to be precise, because this time NASA decided to order directly from the manufacturer. The Swiss company reacted immediately: it added the word "Professional" to the dial and launched an unparalleled advertising campaign that would only be surpassed after the first humans walked on the moon on July 21, 1969.

The watch's popularity grew so much that the movement manufacturer Lemania was faced with bottlenecks and decided to rationalize production of the Caliber 321. Simplification of the design coincided with an improvement of functional reliability. Ever since 1968, the chronograph control has been done via cam lever; the balance frequency was increased from 18,000 to 21,600 half-vibration per hour (3 Hz); and the Breguet mainspring was replaced with a flat spring. At Omega, the new movement was given the number 861. And ever since 1968, Caliber 861 has been put into the model known as the Moon Watch. To this day, the Speedmaster is still the official watch for all "extravehicular activities"—that is, for space trips during which the astronauts take a walk, as it were.

To date, human beings have landed six times on the moon. And the Speedmaster was on each one of those trips.

1945
Automatic Chronometer Centenaire

Case: 18-karat yellow gold, push-down case back, yellow gold Milanaise bracelet, Ø 34 mm
Movement: Caliber 30.10 RA PC JUB, red gold-plated, automatic winding
Remarks: extremely rare anniversary chronometer with hammer automatic movement in celebration of Omega SA's 100th anniversary
Estimated value:: $6,500 ➔

1948
Centenaire

Reference number: OT 2500
Case: 18-karat yellow gold, push-down case back, leather strap, Ø 35 mm
Movement: Caliber 30.10 RA PC, red gold-plated, automatic winding
Remarks: extremely rare anniversary model with hammer automatic movement in celebration of Omega SA's 100th anniversary
Estimated value: $5,200 ➔

1939
Chronograph

Case: 14-karat yellow gold, push-down case back, leather strap, 32 x 40 mm
Movement: Caliber 28.1 CHRO T1, nickel-plated, column-wheel control of chronograph, manual winding
Remarks: one-button chronograph with 30-minute counter in shaped case
Estimated value: $7,800 ↗

1915
Chronograph

Case: silver, push-down case back with hinge, enamel dial, leather strap, Ø 46 mm
Movement: 18''', gold-plated, frosted finish, finely polished and beveled steel chronograph parts, column-wheel control, compensation screw balance, manual winding
Remarks: important early and large pilot's wristwatch, one of the first wristwatches with a chronograph; 15-minute counter
Estimated value: $19,500 ↗

1944
Doctor's Chronograph

Case: stainless steel, push-down case back, leather strap, Ø 38 mm
Movement: Caliber 33.3 CHRO T3, nickel-plated, finely polished steel chronograph parts, column-wheel control, compensation screw balance, manual winding
Remarks: practical doctor's chronograph with a 30-minute counter and pulsometer scale
Estimated value: $13,000 ↗

1944
Doctor's Chronograph

Case: 18-karat yellow gold, push-down case back, leather strap, Ø 38 mm
Movement: Caliber 33.3 CHRO T3, nickel-plated, finely polished steel chronograph parts, column-wheel control, compensation screw balance, manual winding
Remarks: practical doctor's chronograph with a 30-minute counter and pulsometer scale
Estimated value: $13,000 ↗

1944
Military Chronograph

Case: stainless steel, push-down case back, leather strap, Ø 38 mm
Movement: Caliber 33.3 CHRO T2, rhodium-plated, finely polished steel chronograph parts, column-wheel control, compensation screw balance, 17 jewels, manual winding
Remarks: significant chronograph with tachymeter scale in the military design of the early 1940s
Estimated value: $32,500 ➔

1944
Chronograph

Case: stainless steel, screw-down case back, leather strap, Ø 38 mm
Movement: Caliber 33.3 CHRO, gold-plated, frosted finish, finely polished steel chronograph parts, column-wheel control, screw balance, manual winding
Remarks: rare, water-resistant chronograph with a 30-minute counter
Estimated value: $19,500 ↗

1935
Chronograph

Case: stainless steel, push-down case back, leather strap, Ø 37 mm
Movement: Caliber 33.3 CHRO T3, rhodium-plated, column-wheel control of chronograph, manual winding
Remarks: one-button chronograph with 30-minute counter
Estimated value: $10,400 ➔

1938
Seamaster Chronograph

Case: stainless steel, push-down case back, leather strap, Ø 35 mm
Movement: Caliber 321, red gold-plated, column-wheel control of chronograph, manual winding
Remarks: chronograph with 30-minute and 12-hour counters
Estimated value: $9,100 ↗

1953
Royal Air Force Pilot's Watch

Reference number: 2777-1SC
Case: stainless steel, screw-down case back, antimagnetic core, extra long strap, Ø 37 mm
Movement: Caliber 283, red gold-plated, manual winding
Remarks: rare, engraved "6645/101000/6B/5425432/53"; dial printed with "Broad Arrow," "Tritium"; offered in restored condition with excerpt from master registry
Estimated value: $3,250 ➔

1935
Pilot's Watch

Case: stainless steel, push-down case back, rotating bezel, leather strap, Ø 40 mm
Movement: Caliber 26.5 SOB, rhodium-plated, manual winding
Remarks: rare early pilot's watch with rotating bezel
Estimated value: $10,400 ➔

1956
Railmaster

Reference number: 2914-1 SC
Case: stainless steel, screw-down case back, leather strap, Ø 38 mm
Movement: Caliber 284, red gold-plated, 17 jewels, manual winding
Remarks: anti-magnetic men's watch with extra-thick dial
Estimated value: $9,100 ➔

1959
Ranchero

Case: stainless steel, push-down case back, leather strap, Ø 36 mm
Movement: Caliber 267, red gold-plated, manual winding
Remarks: rare men's watch with manually wound movement
Estimated value: $4,290 ➔

1969
De Ville Chronograph

Reference number: 145018
Case: stainless steel, Ø 35 mm, bipartite, screw-down case back, leather strap
Movement: Caliber 861, red gold-plated, fine matte steel components, 17 jewels
Remarks: elegant De Ville chronograph with 30-minute and 12-hour counters; tachymeter scale
Estimated value: $3,900 ➔

1958
Chronograph

Case: 18-karat pink gold-plated, Ø 35 mm, tripartite, push-down case back, leather strap
Movement: Caliber 321, red gold-plated, column-wheel control of chronograph, fine matte steel chronograph components
Remarks: sporty, elegant chronograph with 30-minute and 12-hour counters; tachymeter scale; red gold case
Estimated value: $3,640 ➔

1954
Men's Watch

Case: stainless steel, bipartite, push-down case back, leather strap, 31 x 40 mm
Movement: Caliber 344, red gold-plated, pendulum oscillating weight, index fine adjustment
Remarks: automatic watch in gold design case
Estimated value: $5,200 →

1950
Automatic

Case: 18-karat yellow gold, push-down case back, leather strap, 33 x 33 mm
Movement: Caliber 342, pink gold-plated, glucydur screw balance, automatic winding with an oscillating weight
Remarks: elegant automatic wristwatch with a unique case
Estimated value: $6,500 →

2009
De Ville Co-Axial Chronometer

Case: 18-karat rose gold, screw-down case back with lithograph, leather strap, 34 x 43 mm
Movement: Caliber 2500, rhodium-plated, fausses côtes decoration, coaxial escapement, glucydur screw balance with shock protection, automatic winding, certified chronometer
Remarks: from the Almada Negreiros series, limited to 43 pieces
Estimated value: $7,800 →

1951
Seamaster Automatic Chronometer

Case: stainless steel, screw-down case back, leather strap, Ø 34 mm
Movement: Caliber 342, red gold-plated, regulated in 6 positions, automatic winding
Remarks: chronometer with hammer automatic movement regulated in six positions
Estimated value: $2,210 →

1950
Seamaster Automatic Chronometer

Reference number: 2577/2520 SC
Case: 18-karat yellow gold, bipartite, screw-down case back, leather strap, Ø 34 mm
Movement: Caliber 352, red gold-plated, polished screws, pendulum oscillating weight, index fine adjustment, 17 jewels
Remarks: gold Constellation chronometer
Estimated value: $3,250 →

1953
Seamaster Automatic Chronometer

Case: 14-karat yellow gold, screw-down case back, leather strap, Ø 34 mm
Movement: Caliber 352, red gold-plated, automatic winding
Remarks: men's watch with hammor automatic and chronometer certificate
Estimated value: $ 2,340 →

1971
Seamaster Automatic Officially Certified Chronometer

Case: 18-karat yellow gold, screw-down case back, yellow gold link bracelet, Ø 37 mm
Movement: red gold-plated, automatic winding
Remarks: extremely rare chronometer; this watch was offered in an Omega box
Estimated value: $5,850 →

1958
Seamaster Chronograph

Reference number: OTI 2884
Case: 18-karat yellow gold, push-down case back, leather strap, Ø 35 mm
Movement: Caliber 321, pink gold-plated, finely matted steel chronograph parts, column-wheel control, glucydur screw balance, 17 jewels, manual winding
Remarks: elegant chronograph with a 12-hour and 30-minute counter
Estimated value: $3,900 →

1966
Constellation
Case: 18-karat yellow gold, tripartite, screw-down case back, leather strap, Ø 34 mm
Movement: Caliber 564, red gold-plated, polished screws, winding rotor, 24 jewels, regulated in 6 positions
Remarks: gold Constellation chronometer with date window at 3 o'clock
Estimated value: $2,600 →

1963
Automatic Chronometer Constellation
Case: 18-karat yellow gold, screw-down case back, leather strap, Ø 34 mm
Movement: Caliber 551, red gold-plated, 24 jewels, regulated in 6 positions, automatic winding
Remarks: chronometer with movement regulated in six positions; this watch was offered in its original box and with its original buckle
Estimated value: $3,900 →

1954
Constellation Chronometer De Luxe
Reference number: 2699SC
Case: 18-karat yellow gold, screw-down case back, yellow gold bracelet, Ø 35 mm
Movement: Caliber 2699 SC, pink gold-plated, polished, 17 jewels, regulated in 8 positions, automatic winding, chronometer
Remarks: elegant Constellation Chronometer
Estimated value: $4,550 ↗

1967
Constellation Automatic Chronometer
Reference number: 168.010/11
Case: 18-karat yellow gold, screw-down case back, yellow gold bracelet (Milanese), Ø 36 mm
Movement: Caliber 564, pink gold-plated, polished, 24 jewels, regulated in 6 positions, automatic winding, officially certified chronometer
Remarks: heavy Constellation Chronometer
Estimated value: $5,850 ↗

1962
Constellation Automatic Chronometer
Case: 18-karat red gold, screw-down case back, leather strap, Ø 35 mm
Movement: Caliber 551, red gold-plated, automatic winding
Remarks: rare wristwatch chronometer; this watch was offered in its original box and with its original gold buckle
Estimated value: $3,900 →

1970
Constellation Automatic Chronometer
Case: gold-plated stainless steel, screw-down case back, leather strap, Ø 35 mm
Movement: Caliber 751, red gold-plated, 24 jewels, regulated in 6 positions, automatic winding
Remarks: chronometer with movement regulated in six positions
Estimated value: $1,560 →

1967
Constellation Grand Luxe Calendar Chronometer
Reference number: 14398SC
Case: 18-karat yellow gold, screw-down case back, yellow gold bracelet ("à brique"), Ø 35 mm
Movement: Caliber 561, pink gold-plated, polished, 24 jewels, regulated in 6 positions, automatic winding, chronometer
Remarks: rare Constellation Chronometer; top model of the Omega collection
Estimated value: $6,500 ↗

1958
Constellation Grand Luxe Chronometer
Reference number: 14365SC
Case: 18-karat yellow gold, push-down case back, yellow gold bracelet ("à brique"), Ø 35 mm
Movement: Caliber 501, pink gold-plated, polished, 19 jewels, regulated in 8 positions, automatic winding, chronometer
Remarks: elegant Constellation Chronometer
Estimated value: $7,800 ↗

1968
Automatic Chronometer Constellation

Case: 18-karat yellow gold, screw-down case back, yellow gold link bracelet, Ø 35 mm
Movement: Caliber 564, gold-plated, automatic winding
Remarks: Constellation chronometer; this watch was offered with original box and certificate
Estimated value: $5,850 →

1970
Constellation Automatic Chronometer

Case: gold-plated, stainless steel push-down case back, leather strap, Ø 38 mm
Movement: Caliber 751, red gold-plated, 24 jewels, regulated in 6 positions, automatic winding
Remarks: extra-large Constellation chronometer with movement regulated in six positions
Estimated value: $1,300 →

1973
De Ville

Reference number: 1011
Case: silver, push-down case back, crown with lapis lazuli cabochon, bracelet with ivory inlays, 36 x 24 mm
Movement: pink gold-plated, polished, glucydur balance, manual winding
Remarks: cult wristwatch from the early 1970s
Estimated value: $3,900 →

1973
De Ville

Reference number: 1011
Case and band: silver and ivory, push-down case back, crown with lapis lazuli cabochon, 24 x 33 mm
Movement: Caliber 825, pink gold-plated, polished, manual winding
Remarks: extremely rare men's wristwatch in the exalted style of the early 1970s
Estimated value: $2,600 →

1957
Ranchero

Case: 18-karat yellow gold, push-down case back, leather strap, Ø 35 mm
Movement: Caliber 267, red gold-plated, manual winding
Remarks: rare Omega Ranchero in yellow gold with a white dial
Estimated value: $3,900 →

1954
Men's Watch with Enamel Dial

Reference number: 2686
Case: 18-karat yellow gold, push-down case back, leather strap, Ø 34 mm
Movement: Caliber 283, pink gold-plated, glucydur balance, 17 jewels, manual winding
Remarks: unusual wristwatch with a rare cloisonné enamel dial
Estimated value: $10,400 →

1953
Cosmic

Reference number: 3944
Case: 14-karat yellow gold, push-down case back, leather strap, 33 x 44 mm
Movement: Caliber 381, pink gold-plated, polished, glucydur balance, 17 jewels, manual winding
Remarks: rare men's wristwatch with calendar and moon phase
Estimated value: $9,100 →

1966
Chronograph

Case: 18-karat rose gold, push-down case back, leather strap, Ø 35 mm
Movement: Caliber 321, pink gold-plated, finely matted steel chronograph parts, column-wheel control, glucydur screw balance, manual winding
Remarks: elegant chronograph with a 12-hour and 30-minute counter
Estimated value: $5,200 →

1958

Speedmaster "Broad Arrow"

Reference number: CK2915
Case: stainless steel, screw-down case back, stainless steel bracelet, Ø 39 mm
Movement: Caliber 321, pink gold-plated, finely polished steel chronograph parts, column-wheel control, glucydur screw balance with shock protection, manual winding
Remarks: important chronograph with a 12-hour and 30-minute counter; the first Speedmaster: with an original Omega strap
Estimated value: $52,000 ➔

1970

Automatic Seamaster 300

Case: stainless steel, screw-down case back, rotating bezel, leather strap, Ø 42 mm
Movement: Caliber 565, red gold-plated, automatic winding
Remarks: Seamaster with additional transparent case back
Estimated value: $3,250 ➔

1980

Speedmaster Professional Chronograph

Case: stainless steel, screw-down case back, leather strap, Ø 42 mm
Movement: Caliber 861, red gold-plated, manual winding
Remarks: chronograph with 30-minute and 12-hour counters
Estimated value: $3,120 ➔

1980

Speedmaster Professional Chronograph

Case: stainless steel, screw-down case back, leather strap, Ø 42 mm
Movement: Caliber 866, red gold-plated, manual winding
Remarks: chronograph with 30-minute and 12-hour counters; moon phase and date
Estimated value: $3,900 ➔

1970

Speedmaster Professional

Reference number: 145022
Case: 18-karat yellow gold, screw-down case back, yellow gold link bracelet, Ø 42 mm
Movement: Caliber 861, red gold-plated, manual winding
Remarks: gold Speedmaster chronograph; limited to 1,014 pieces (this watch is engraved with number 347/1014); solid gold
Estimated value: $11,700 ➔

1968

Seamaster Chronograph

Case: 18-karat yellow gold, screw-down case back, leather strap, Ø 37 mm
Movement: Caliber 321, gold-plated, column-wheel control of chronograph, manual winding
Remarks: rare gold chronograph with 30-minute and 12-hour counters
Estimated value: $8,450 ➚

1986

Seamaster Soccer Timer

Reference number: 145.019-69
Case: stainless steel, screw-down case back, crown-turned "roulette" hour disc, leather strap, Ø 44 mm
Movement: Caliber 861, pink gold-plated, finely polished steel chronograph parts, glucydur balance, 17 jewels, manual winding
Remarks: sporty chronograph with a 12-hour and 30- or 45-minute counter
Estimated value: $5,200 ➚

1971

Seamaster Chronograph

Case: stainless steel, screw-down case back, leather strap, 41 x 46 mm
Movement: Caliber 861, red gold-plated, manual winding
Remarks: chronograph with 30-minute and 12-hour counters; subsidiary dial scale additionally extended to 45 minutes
Estimated value: $2,990 ➔

1970
Seamaster Chronostop

Case: stainless steel, screw-down case back, leather strap, 41 x 47 mm
Movement: Caliber 865, red gold-plated, manual winding
Remarks: simple chronograph with rotating inner bezel (flange)
Estimated value: $3,250 ➔

1970
Seamaster Soccer Timer

Reference number: 145.019
Case: stainless steel, screw-down case back, crown-controlled scale disc, stainless steel bracelet, Ø 44 mm
Movement: Caliber 861, pink gold-plated, finely polished steel chronograph parts, glucydur balance, 17 jewels, manual winding
Remarks: sporty chronograph with a 12-hour and 30- or 45-minute counter; rare "soccer" dial with halftime display
Estimated value: $3,640 ➔

1971
Seamaster Professional 1000m/3300ft Automatic

Case: stainless steel, screw-down case back, rotating bezel, screw-in crown left, stainless steel bracelet, Ø 45 mm
Movement: Caliber 1002, pink gold plated, glucydur balance, automatic winding
Remarks: professional deep-sea diver's watch with date
Estimated value: $7,800 ➔

1969
Bullhead Chronograph

Case: stainless steel, screw-down case back, crown-controlled scale ring, leather strap, 43 x 42 mm
Movement: Caliber 930, pink gold-plated, finely polished steel chronograph parts, glucydur balance, manual winding
Remarks: asymmetrical chronograph in a design typical of the late 1960s
Estimated value: $5,850 ➚

1977
Seamaster Automatic Chronograph with Lemania Engraving

Case: 18-karat yellow gold, screw-down case back, leather strap, 38 x 43 mm
Movement: Caliber 1040, gold-plated, automatic winding
Remarks: automatic chronograph with sweep minute counter; 12-hour counter; 24-hour display; according to engraving, this watch was a gift from Lemania to an employee: "Lemania à Albert Piguet 1934-1977"
Estimated value: $6,500 ➔

1974
Automatic Chronograph FIFA

Case: stainless steel, screw-down case back, leather strap, 40 x 43 mm
Movement: Caliber 1045, red gold-plated, automatic winding
Remarks: special edition chronograph for world soccer organization FIFA
Estimated value: $3,120 ➔

1975
Seamaster Professional 600m/2000ft Automatic

Case: stainless steel, lockable rotating bezel, screw-in crown left, screw-down case back, stainless steel bracelet, 55 x 45 mm
Movement: Caliber 1002, pink gold-plated, glucydur balance, automatic winding
Remarks: 2nd generation professional deep-sea watch PLOPROF ("plongeurs professionels")
Estimated value: $9,100 ➚

1968
Chronostop

Case: stainless steel, screw-down case back, leather strap, 35 x 39 mm
Movement: Caliber 865, red gold-plated, manual winding
Remarks: simple chronograph with second counter
Estimated value: $780 ➔

1970
Seamaster Chronograph Darth Vader

Case: blackened stainless steel, screw-down case back, leather strap, 44 x 46 mm
Movement: Caliber 861, red gold-plated, manual winding
Remarks: chronograph with black case; 30-minute and 12-hour counters
Estimated value: $5,200 ➔

1973
Speedmaster Professional Anakin Skywalker

Case: stainless steel, screw-down case back, stainless steel link bracelet, 45 x 51 mm
Movement: Caliber 861, red gold-plated, manual winding
Remarks: heavy chronograph with 30-minute and 12-hour counters
Estimated value: $5,850 ➔

1975
Flightmaster

Case: 18-karat yellow gold, 43 x 52 mm, bipartite, screw-down case back, gold link bracelet
Movement: Caliber 910, red gold-plated, fine matte steel chronograph components
Remarks: extremely rare, heavy Flightmaster in a solid gold case (217 grams); 30-minute and 12-hour counters; 24-hour and second time zone displays; this watch was offered with its original gold link bracelet
Estimated value: $19,500 ➚

1975
Flightmaster

Case: stainless steel, screw-down case back, stainless steel link bracelet, Ø 43 mm
Movement: Caliber 911, red gold-plated, manual winding
Remarks: chronograph with 30-minute and 12-hour counters; second time zone; inner rotating bezel (flange)
Estimated value: $3,900 ➔

1975
Speedmaster Professional Mk IV

Case: stainless steel, screw-down case back, stainless steel link bracelet, Ø 42 mm
Movement: Caliber 1041, red gold-plated, automatic winding
Remarks: heavy automatic chronograph with sweep minute counter; 12-hour counter; 24-hour display
Estimated value: $1,950 ➔

1972
Speedmaster Professional Mk II

Case: stainless steel, screw-down case back, stainless steel link bracelet, Ø 42 mm
Movement: Caliber 861, red gold-plated, manual winding
Remarks: heavy chronograph with 30-minute and 12-hour counters
Estimated value: $1,690 ➔

1977
Speedmaster 125

Case: stainless steel, screw-down case back, stainless steel link bracelet, 42 x 51 mm
Movement: Caliber 1041, red gold-plated, 22 jewels, regulated in 6 positions, automatic winding
Remarks: special edition model for Omega's 125th anniversary; chronograph with sweep minute counter; 12-hour counter; 24-hour display
Estimated value: $2,600 ➔

1980
Speedmaster Professional Mk V

Case: stainless steel, stainless steel link bracelet, Ø 43 mm
Movement: Caliber 1045, red gold-plated, automatic winding
Remarks: heavy automatic chronograph with sweep minute counter; 12-hour counter; 24-hour display
Estimated value: $3,250 ➔

1975

Seamster Chronograph Automatic

Case: stainless steel, screw-down case back, stainless steel link bracelet, 42 x 48 mm
Movement: Caliber 1040, red gold-plated, automatic winding
Remarks: heavy automatic chronograph with sweep minute counter; 12-hour counter; 24-hour display
Estimated value: $1,430 ➔

1978

Speedmaster Chronograph Automatic

Case: stainless steel, screw-down case back, steel link bracelet, Ø 40 mm
Movement: Caliber 1045, red gold-plated, automatic winding
Remarks: automatic chronograph with sweep minute counter; 12-hour counter; 24-hour display
Estimated value: $1,690 ➔

1975

Seamaster Chronograph Automatic

Case: stainless steel, screw-down case back, leather strap, 41 x 48 mm
Movement: Caliber 1040, red gold-plated, automatic winding
Remarks: large automatic chronograph with sweep minute counter; 12-hour counter; 24-hour display
Estimated value: $1,560 ➔

1972

Memomatic

Reference number: 166.071
Case: stainless steel, screw-down case back, leather strap, Ø 42 mm
Movement: Caliber 980, pink gold-plated, 19 jewels, gongs, automatic winding
Remarks: wristwatch alarm with minute setting of wake-up in the modern casing of the Seamaster
Estimated value: $1,820 ➔

1975

Marine Chronometer Constellation

Reference number: 1980074
Case: stainless steel / yellow gold, 32 x 49 mm, tripartite, screw-down case back, link bracelet
Movement: Megaquartz Caliber 1511, red gold-plated, polished screws, 13 jewels
Remarks: this was the first wristwatch to fulfill all demands of a certified marine chronometer; offered with chronometer certificate and rate report of the Besançon observatory
Estimated value: $3,640 ➚

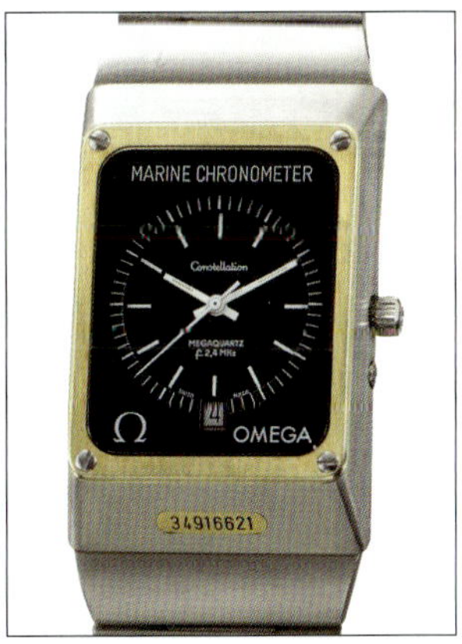

1972

Stardust Constellation

Case: 18-karat yellow gold, 32 x 48 mm, bipartite, screw-down case back, leather strap
Movement: Megaquartz Caliber 1510, electromechanical
Remarks: extremely rare heavy Constellation with precision quartz movement Megaquartz 2.4 MHz; dial crafted in aventurine to look like starry sky; this watch is number 409 from a special series of only 1,000 pieces
Estimated value: $ 3,900 ➔

1956

Seamaster XVI

Case: 18-karat red gold, push-down case back, leather strap, Ø 34 mm
Movement: Caliber 471, red gold-plated, automatic winding
Remarks: rare red gold Seamaster XVI in honor of the sixteenth Olympic Games in Melbourne; engraved Olympic cross on case back
Estimated value: $4,550 ➔

2005

"The Petrograd Watch"

Case: 18-karat rose gold, screwed-down case back, leather strap, 32 x 43 mm
Movement: Caliber 2200, rhodium-plated, fausses côtes decoration, glucydur screw balance with shock protection, automatic winding
Remarks: from the Omega Museum Collection, Series 4, limited to 1,915 pieces
Estimated value: $7,150 ➘

OFFICINE PANERAI

When Giovanni Panerai founded his little workshop in 1860 in a building on the Ponte alle Grazie, a bridge lined with buildings in Florence, he laid the cornerstone for one of the most exceptional watch companies.

Already in 1871 he had to move his workshop from the Ponte alle Grazie, and after a number of moves, house number 3 on Piazza San Giovanni became his final domicile. The little workshop also represented the first watchmaking school in Florence.

Orologeria Svizzeria (Swiss Watchmaking), as Panerai named his shop, sold and repaired watches imported from Switzerland. Under the same roof and clearly separated from the retail shop, Panerai began to manufacture precision instruments.

The rise of Officine Panerai (literally: "Panerai's workshops") to a position of highly specialized supplier of the Italian navy took place under Giovanni's grandson Guido. He put special attention on good precision rates, and thanks to this, Panerai was soon supplying both the Italian railways and officers of the Royal Navy.

After that, the company became an important source for the Italian navy, but not as one might expect for watches, but rather for depth gauges and marine compasses for the wrist as well as other equipment.

Thanks to the luminous substance Radiomir invented in 1910—a mixture of zinc sulfate and radium bromide—the Panerai dials were especially legible deep under water and at night. Panerai also used this to advantage for the company's watches, and thus Guido received the contract to produce the first wristwatches for combat divers, which he also called Radiomir. The first prototypes were produced by Rolex, though the following watches housed in cases according to the Oyster principle were made by Panerai. Thanks to the minimal production numbers, Panerai purchased movements from Aegler, Rolex Manufacture in Biel, and Angelus.

In 1936, when the first ten watches had been delivered, the Panerai Radiomir became a secret piece of navy equipment. The Radiomir was followed by the Mare Nostrum chronograph, which hardly made it past the prototype stage, and the Luminor with its striking crown protection. Only long after World War II was over did Panerai seek a patent for it, in 1955. Its easy-to-use bar presses the crown into a cork gasket, ensuring absolute water resistance up to 200 meters. The bar is simple to open, and an unsecured crown is immediately recognizable in this position, thus reducing the risk of damage.

After World War II, the Radiomir and the Luminor continued to be manufactured practically unchanged, at first for the armed services and then, starting in 1993, in a limited edition of about 1,000 pieces for the civilian market. This included a Slytech edition for the American actor Sylvester (Sly) Stallone.

The striking visuals of the professional diving instruments, their early tie to cult brand Rolex, inclusion of technology related to that of Rolex, and the fact that between 1938 and 1994 only about 350 Panerai watches were manufactured, secured the brand complete success at its relaunch under the auspices of the Richemont Luxury Group in 1997. Vintage Panerai watches originating in the early days also continue to profit from these elements—particularly at auction.

Founded:
1860 by Giovanni Panerai in Florence

Headquarters:
Florence, today Neuchâtel

Status:
active without interruption (today Richemont Luxury Group)

Specialty:
diving watches, marine chronographs, railway watches, measuring instruments, dashboard clocks

Major inventions:
- 1910 Radiomir luminescent mass
- 1936 first diving watches for combat frogmen
- 1942 crown protection with integrated lever (patented in 1955)
- 1949 Luminor luminescent mass

Leading models:
Radiomir, Luminox

ca. 1950
Luminor

Case: stainless steel, screw-down case back, leather strap, Ø 47 mm
Movement: rhodium-plated, côtes de Genève, jewels set in chatons, manual winding
Remarks: combat diver's watch of the Italian navy; typical crown protection
Estimated value: $78,000 ↗

ca. 1944
Radiomir

Reference number: 3646/Type C
Case: stainless steel, screw-down case back, leather strap, Ø 46 mm
Movement: Caliber Cortébert 16 (Rolex 618/Type 1), nickel-plated, fausses côtes decoration, jewels set in chatons, screw balance, 17 jewels, manual winding
Remarks: important combat swimmer watch of the German navy (owner: Fred Richard), only 50 pieces of this type are known
Estimated value: $78,000 ↗

ca. 1940
Radiomir

Reference number: 3646/Type B
Case: stainless steel, screw-down case back, leather strap, Ø 46 mm
Movement: Caliber Cortébert 16 (Rolex 618/Type 1), nickel-plated, fausses côtes decoration, jewels set in chatons, screw balance, 17 jewels, manual winding
Remarks: significant combat swimmer watch of the Italian navy, only 13 pieces of this type are known
Estimated value: $91,000 ↗

ca. 1950
Luminor Marina Militare

Case: stainless steel, screw-down case back, leather strap, Ø 47 mm
Movement: rhodium-plated, côtes de Genève, jewels set in chatons, manual winding
Remarks: combat diver's watch of the Italian navy with rare words "Marina Militare" printed on dial
Estimated value: $117,000 ↗

archive photo

ca. 1944
Radiomir

Reference number: 3646/Type D
Case: stainless steel, screw-down case back, leather strap, Ø 46 mm
Movement: Caliber Cortébert 16 (Rolex 618/Type 1), nickel-plated, fausses côtes decoration, jewels set in chatons, screw balance, 17 jewels, manual winding
Remarks: major combat swimmer watch of the German navy with an original strap and lead gasket (owner: Bernhard Dornsiepen), only 52 pieces of this type are known
Estimated value: $104,000 ↗

ca. 1938
Radiomir Brevettato

Reference number: 3646
Case: stainless steel, screw-down case back, leather strap, Ø 47 mm
Movement: rhodium-plated, côtes de Genève, jewels set in chatons, manual winding
Remarks: important combat diver's watch of the Italian navy; this watch is a prototype only made for presentation purposes; case back signed "Oyster Watch Geneva Swiss"
Estimated value: $91,000 ↗

ca. 1953
Radiomir

Reference number: 6152
Case: stainless steel, screw-down case back, leather strap, Ø 47 mm
Movement: Caliber Rolex 618/Type 2, nickel-plated, fausses côtes decoration, jewels set in chatons, screw balance, 17 jewels, manual winding
Remarks: significant and very rare marine diver's watch, fewer than 10 pieces of this type are known
Estimated value: $195,000 ↗

ca. 1956
Luminor Egittica

Case: stainless steel, rotating bezel, screw-down case back, leather strap, Ø 47 mm
Movement: Angelus 8-day movement, rhodium-plated, manual winding
Remarks: combat diver's watch for the Egyptian navy with eight-day movement and rotating bezel; only about 50 pieces produced
Estimated value: $182,000 ↗

archive photo

PATEK PHILIPPE

the Master Watchmakers of Geneva

PATEK PHILIPPE

The history of Patek Philippe began with a Polish count, Antoine Norbert de Patek (1812–1877), who spent time in political exile in Geneva, thereby discovering a love for fine watchmaking. Together with countryman François Czapek, a talented watchmaker, the former officer founded Patek, Czapek und Cie in 1839.

With the aid of six employees they produced about 200 watches per year. The aesthetic and technical qualities of these timepieces secured them an enviable reputation right from the beginning of the young company's history.

Patek first heard of a watchmaker named Jean-Adrien Philippe (1815–1894) at an exhibition in Paris where he was offering his wares. This watchmaker had made a flat pocket watch that could be both set and wound by the crown—an improvement of barely imaginable dimensions today.

Since Patek's contract with Czapek had run out, he offered Philippe a partnership, and on May 15, 1845, he renamed his company Patek & Co. Since the success of his watches could be attributed to the genius-like watchmaker Philippe, he then added the latter's name to that of the company for Patek Philippe & Co. starting in 1851.

The duo had great influence on the development of the watch industry as the sights of both these men were set high: they wanted to make the best watches in the world. Philippe turned out to be a creative visionary, pushing the industrialization and construction of new tooling forward. He also understood how to introduce his complicated watches to a wide audience, presenting his products at numerous exhibitions. In London in 1851, Queen Victoria purchased a gold pocket watch with Philippe's famous crown winding. Many crowned heads and grand personalities followed her example.

Patek Philippe created its first wristwatch in 1868 for the Hungarian Countess Kocewicz—and this may well have been the very first wristwatch in the history of Swiss watchmaking.

The passing of Antoine de Patek in March 1877 did not slow Jean-Adrien Philippe's need to innovate. In 1889 he designed a watch movement containing a "perpetual" date, the display of which needed no correction whatsoever, either for the differing lengths of the months or for the leap years.

When in 1894 Philippe also passed away, his son-in-law Joseph-Antoine Benassy-Philippe took over the management of the endeavor and turned it into a joint stock company seven years later.

Even though the Philippe family had to sell Patek Philippe as a result of the great economic crisis of 1929, today the company is still in the hands of a single family: the Stern brothers, until that point the exclusive dial suppliers to Patek Philippe, purchased the *manufacture* in 1932 and took it to a new level. Under their aegis, the finest wristwatches ever seen were made in Geneva, containing many horological complications—from the perpetual calendar to the chronograph and the tourbillon. The brand's simple, timeless models of extraordinarily exquisite craftsmanship from this era also enjoy a very special place in the world of collecting. The most expensive watch ever to be sold at auction in Germany is—of course—a Patek Philippe.

Founded:
1845 by Antoine Norbert de Patek and Jean-Adrien Philippe

Headquarters:
Geneva

Status:
active without interruption

Specialty:
high-end wristwatches, chronometers, chronographs (with split seconds), perpetual calendars, repeaters, tourbillons, and *grandes complications*

Major inventions:
- 1842 invention of the crown winder
- 1845 pocket watch with minute repeater
- 1863 mainspring with bridle
- 1868 golden bracelet with integrated watch
- 1869 perpetual calendar
- 1928 Opus Technicus with twenty-four complications
- 1932 introduction of the first Calatrava
- 1989 Caliber 89 with thirty-three complications

Leading models:
Calatrava, Gondolo World Time, Nautilus

1913

Chronometro Gondolo

Case: 18-karat yellow gold, push-down case back, leather strap, 26 x 44 mm
Movement: Caliber 10, frosted finish, gold-plated, "moustache" compensating pallets, manual winding
Remarks: one of Patek Philippe's earliest wristwatches from 1913; this watch was offered with a Patek Philippe certificate
Estimated value: $52,000 ➔

1918

Chronometro Gondolo

Case: 18-karat yellow gold, push-down case back, leather strap, 31 x 48 mm
Movement: Caliber 12, gold-plated, "moustache" compensating pallets, manual winding
Remarks: one of Patek Philippe's earliest wristwatches from 1918; tonneau-shaped case; this watch was offered with a gold buckle and a Patek Philippe certificate
Estimated value: $39,000 ➔

1920

Men's Watch

Case: 18-karat yellow gold, push-down case back, leather strap, 26 x 43 mm
Movement: Caliber 10, frosted finish, gold-plated, "moustache" compensating pallets, manual winding
Remarks: one of Patek Philippe's earliest wristwatches from 1920; exceptional case with guilloché dial; this watch was offered with gold buckle and certificate
Estimated value: $16,900 ➚

1928

Men's Watch Art Deco

Case: 18-karat white gold, push-down case back, leather strap, 25 x 40 mm
Movement: Caliber 10, rhodium-plated, côtes de Genève, "moustache" compensating pallets, 18 jewels, regulated in 8 positions, manual winding
Remarks: rare Art Deco men's watch in a rectangular case; hand-engraved case
Estimated value: $32,500 ➔

1928

Men's Watch

Case: 18-karat white gold, push-down case back, leather strap, 24 x 38 mm
Movement: gold-plated, manual winding
Remarks: elegant white gold men's watch in rectangular case
Estimated value: $11,700 ➔

1936

Men's Watch

Reference number: 417
Case: platinum, bipartite, push-down case back, leather strap, 20 x 36 mm
Movement: Caliber 9-90, rhodium-plated, fausses côtes decoration, mirror-polished screws, shaped movement, 18 jewels, regulated in 8 positions
Remarks: rare early platinum watch in a stepped case; engraving "S.S. Magoffin" on case back
Estimated value: $32,500 ➚

1935

Men's Watch

Case: 18-karat yellow gold, bipartite, push-down case back with hinge, leather strap, 24 x 40 mm
Movement: Caliber 9-90, rhodium-plated, fausses côtes decoration, mirror-polished screws, shaped movement
Remarks: extremely rare men's watch in a lavish rectangular case
Estimated value: $23,400 ➔

1937

Men's Watch

Reference number: 520
Case: 18-karat yellow gold, bipartite, push-down case back, leather strap, 24 x 40 mm
Movement: Caliber 9-90, rhodium-plated, fausses côtes decoration, mirror-polished screws, shaped movement
Remarks: elegant men's watch in a rectangular case; this watch was offered with an original Patek Philippe buckle and an excerpt from the master registry
Estimated value: $13,000 ➔

PATEK PHILIPPE NAUTILUS

BELOVED BLACK SHEEP

Reference 3700/1 never really managed to fit in with the high-end Geneva *manufacture* of Patek Philippe. A stainless steel watch with a link bracelet, on top of it all water-resistant and particularly robustly designed—who would have ever thought that a notorious specialist in complicated watches and fine, ultra-thin men's watches, would ever come up with something like that? Well, those who could afford the watch saw it coming. One could argue that it was actually Sean Connery as James Bond in *Dr. No* who helped engineer the social breakthrough of sports watches made of steel (by the way, it was a Rolex Submariner back then). On the other hand, no one denies the fact that it was the surprise of the Basel Samples Fair of 1972, the extremely expensive steel Royal Oak by Audemars Piguet, which launched the trend of luxury sports watches.

The name Gérald Genta, the designer of the Royal Oak, was suddenly on everyone's lips. He designed new models for IWC and Vacheron Constantin—and for Patek Philippe. The watch was called the Nautilus, and Gérald Genta also based its design on the bull's-eye principle, only this time the flange screws were not visible, but rather like a bull's-eye that opens using a lateral hinge on a pivot. And following the style of the company, the dimensions were all a little lighter, less edgy; the dial was a little more elegant; the fit of the bracelet more snug. And, naturally, the watch was a whole lot more expensive than the Royal Oak. The original Nautilus from 1976 had a Jaeger-LeCoultre automatic in it (like the Royal Oak). The new Patek Philippe automatic movement was only ready for the next model, reference 3800/1. The Nautilus remains expensive to this day. The Reference 3700/1, which is now approaching its fiftieth year, costs well over $25,000.

1938
Men's Watch

Case: stainless steel, push-down case back, leather strap, 23 x 40 mm
Movement: Caliber 9-90, rhodium-plated, côtes de Genève, 18 jewels, regulated in 8 positions, manual winding
Remarks: simple men's watch in rectangular stainless steel case
Estimated value: $20,800 →

1938
Men's Watch

Reference number: 491
Case: 18-karat yellow gold, push-down case back, leather strap, 21 x 43 mm
Movement: Caliber 9-90, rhodium-plated, côtes de Genève, 18 jewels, regulated in 8 positions, manual winding
Remarks: elegant early men's watch in rectangular case; this watch was offered with an excerpt from the Patek Philippe master registry, a gold buckle, and in its original box
Estimated value: $18,200 →

1940
Men's Watch

Case: 18-karat yellow gold/red gold, push-down case back, leather strap, 21 x 37 mm
Movement: Caliber 9-90, rhodium-plated, côtes de Genève, 18 jewels, regulated in 8 positions, manual winding
Remarks: extremely rare red and white gold men's watch; the watch was sold by Tiffany & Co. in New York
Estimated value: $19,500 →

1937
Top Hat

Case: 18-karat red gold, push-down case back, gold link bracelet
Movement: Caliber 8L, nickel-plated, côtes de Genève, 18 jewels, regulated in 8 positions, manual winding
Remarks: rare rectangular men's watch whose shape is reminiscent of a top hat
Estimated value: $7,800 ↗

1938
Top Hat

Case: 18-karat red gold, push-down case back, leather strap, 25 x 36 mm
Movement: Caliber 9-90, rhodium-plated, côtes de Genève, manual winding
Remarks: rare rectangular men's watch whose shape is reminiscent of a top hat
Estimated value: $16,900 ↗

1947
Men's Watch

Case: 18-karat yellow gold, push-down case back, leather strap, 23 x 36 mm
Movement: Caliber 9-90, rhodium-plated, côtes de Genève, Seal of Geneva, manual winding
Remarks: elegant men's watch with subsidiary seconds in a rectangular case with rolling-pin shaped lugs
Estimated value: $3,900 ↗

1947
Men's Watch

Case: 18-karat red gold, leather strap, 26 x 37 mm
Movement: Caliber 10-200, rhodium-plated, côtes de Genève, manual winding
Remarks: rare men's watch with patented dust and moisture-protected case; the case is opened by a slide on the side
Estimated value: $5,200 ↗

1945
Shark

Reference number: 2554
Case: 18-karat white gold, bipartite, push-down case back, leather strap, 28 x 43 mm
Movement: Caliber 9-90, rhodium-plated, fausses côtes, mirror-polished screws, shaped, 18 jewels, regulated in 8 positions
Remarks: extremely rare men's watch; Ref. 2554 served as the basis for Ref. 5100 with ten days' power reserve in 2000; this watch was offered with a Patek Philippe buckle, an excerpt from the master registry, and its original box
Estimated value: $39,000 ↗

1940
Men's Watch

Reference number: 1530
Case: 18-karat red gold, push-down case back, leather strap, 25 x 36 mm
Movement: Caliber 9-90, rhodium-plated, côtes de Genève, Seal of Geneva, 18 jewels, regulated in 8 positions, manual winding
Remarks: red gold men's watch in a rectangular case with double signature; this watch was sold by Tiffany & Co.
Estimated value: $4,550 ↗

1952
Men's Watch

Reference number: 2404
Case: 18-karat yellow gold, push-down case back, leather strap, 24 x 37 mm
Movement: Caliber 9-90, rhodium-plated, côtes de Genève, Seal of Geneva, 18 jewels, regulated in 8 positions, manual winding
Remarks: men's watch with subsidiary seconds; this watch has an exceptional case with remarkable lugs and a strongly domed crystal
Estimated value: $5,850 ↗

1950
Men's Watch

Reference number: 1560
Case: 18-karat yellow gold, push-down case back, leather strap, 22 x 38 mm
Movement: Caliber 9-90, rhodium-plated, côtes de Genève, Seal of Geneva, manual winding
Remarks: elegant men's watch in rectangular case
Estimated value: $4,550 ↗

1960
Men's Watch

Reference number: 2476
Case: 18-karat white gold, bipartite, push-down case back, leather strap, 26 x 39 mm
Movement: Caliber 9-90, rhodium-plated, fausses côtes decoration, mirror-polished screws, shaped movement
Remarks: fine elegant men's watch in white gold shaped case; this watch was offered with an excerpt from the Patek Philippe master registry
Estimated value: $32,500 ↗

1953
Men's Watch

Reference number: 2529
Case: 18-karat yellow gold, bipartite, push-down case back, leather strap, 28 x 36 mm
Movement: Caliber 10-200, rhodium-plated, fausses côtes decoration, polished screws
Remarks: elegant men's watch with stepped center part; this watch was offered with an 18-karat gold buckle and an excerpt from the Patek Philippe master registry
Estimated value: $5,200 ↗

1938
Asymmetric Men's Watch

Reference number: 1401
Case: 18-karat yellow gold, push-down case back, leather strap, 20 x 38 mm
Movement: Caliber 9-90, rhodium-plated, côtes de Genève, Seal of Geneva, 18 jewels, regulated in 8 positions, manual winding
Remarks: extremely rare men's watch with asymmetrical case that lies on the wrist in a slightly slanted manner
Estimated value: $26,000 ↗

1940
Men's Watch

Reference number: 1480
Case: 18-karat rose gold, push-down case back, leather strap, 22 x 40 mm
Movement: Caliber 9-90, rhodium-plated, côtes de Genève, Seal of Geneva, 18 jewels, regulated in 8 positions, manual winding
Remarks: elegant men's watch with remarkably shaped, fluted lugs
Estimated value: $19,500 ↗

1955
Hour Glass

Reference number: 1593
Case: 18-karat yellow gold, push-down case back, leather strap, 25 x 41 mm
Movement: Caliber 9-90, rhodium-plated, côtes de Genève, Seal of Geneva, 18 jewels, regulated in 8 positions, manual winding
Remarks: extremely rare men's watch with strongly fitted case reminiscent of an hourglass; this watch was offered in its original box and with a gold buckle
Estimated value: $19,500 ↗

1951
Men's Wristwatch

Reference number: 2442
Case: 18-karat yellow gold, push-down case back, leather strap, 27 x 43 mm
Movement: Caliber 9-90, rhodium-plated, côtes de Genève, Seal of Geneva, manual winding
Remarks: extremely rare men's watch strongly fitted case; double signature of Patek Philippe and Gübelin on dial; this watch was offered with a Patek Philippe certificate
Estimated value: $45,500 ↗

1949
Men's Watch

Reference number: 2456
Case: 18-karat yellow gold, push-down case back, leather strap, 23 x 39 mm
Movement: Caliber 9-90, rhodium-plated, côtes de Genève, 18 jewels, manual winding
Remarks: elegant men's watch in unusually designed rectangular case; this watch was offered with an excerpt from the master registry, a gold clasp, and its original box
Estimated value: $7,800 ↗

1965
Men's Watch

Reference number: 1588
Case: 18-karat yellow gold, push-down case back, leather strap, 23 x 37 mm
Movement: Caliber 9-90, rhodium-plated, côtes de Genève, manual winding
Remarks: simple men's watch in rectangular case
Estimated value: $7,800 ↗

1950
Men's Watch

Reference number: 2434
Case: 18-karat yellow gold, push-down case back, leather strap, 25 x 40 mm
Movement: Caliber 9-90, rhodium-plated, côtes de Genève, 18 jewels, regulated in 8 positions, manual winding
Remarks: elegant men's watch in rectangular case; this watch was offered with an excerpt from the master registry
Estimated value: $5,200 ↗

1954
Men's Watch

Reference number: 2472
Case: 18-karat yellow gold, push-down case back, leather strap, 27 x 36 mm
Movement: Caliber 10-200, rhodium plated, côtes de Genève, manual winding
Remarks: fine red gold men's watch; this watch was offered in its original box
Estimated value: $7,800 ↗

1945
Men's Watch

Case: 18-karat yellow gold, push-down case back, leather strap, 20 x 42 mm
Movement: Caliber 9-90, rhodium-plated, côtes de Genève, manual winding
Remarks: elegant men's watch in rectangular case
Estimated value: $9,100 ↗

1935
Asymmétrique

Reference number: 497
Case: 18-karat yellow gold, push-down case back, leather strap, 25 x 37 mm
Movement: Caliber 91, rhodium-plated, côtes de Genève, Seal of Geneva, 18 jewels, regulated in 8 positions, manual winding
Remarks: rare men's watch with exceptional case; the design of the strap lugs dominate the shape of the case
Estimated value: $13,000 ↗

1955
Men's Watch

Reference number: 2469
Case: 18-karat yellow gold, push-down case back, leather strap, 27 x 38 mm
Movement: Caliber 9-90, rhodium-plated, côtes de Genève, 18 jewels, regulated in 8 positions, manual winding
Remarks: men's watch in a tonneau-shaped case with subsidiary seconds
Estimated value: $6,500 ↗

1941
Calatrava

Reference number: 96
Case: stainless steel, push-down case back, leather strap, Ø 31 mm
Movement: rhodium-plated, fausses côtes decoration, glucydur screw balance, manual winding
Remarks: very fine, rare, elegant Geneva men's watch
Estimated value: $11,300 →

1949
Calatrava

Reference number: 565
Case: stainless steel, screw-down case back, additional movement protection cap, leather strap
Movement: Caliber 12-120, rhodium-plated, fausses côtes, Seal of Geneva, 18 jewels, regulated in 8 positions
Remarks: elegant men's watch in stainless steel; offered with original Patek Philippe buckle and excerpt from master registry
Estimated value: $26,000 ↗

1938
Calatrava

Case: stainless steel, push-down case back, leather strap, Ø 30 mm
Movement: Caliber 12-120, rhodium-plated, côtes de Genève, 18 jewels, regulated in 8 positions, manual winding
Remarks: elegant men's watch with subsidiary seconds
Estimated value: $7,800 ↗

1949
Calatrava

Case: 18-karat red gold, push-down case back, leather strap, Ø 30 mm
Movement: Caliber 12SC, rhodium-plated, côtes de Genève, manual winding
Remarks: rare men's watch
Estimated value: $6,500 ↗

1935
Calatrava

Reference number: 35
Case: stainless steel, tripartite, push-down case back, leather strap, Ø 31 mm
Movement: 12 Lines, rhodium-plated, fausses côtes decoration, polished screws, Lépine wolf-toothed winding wheels
Remarks: elegant Calatrava with subsidiary seconds in stainless steel case; this watch was offered with a Patek Philippe buckle
Estimated value: $8,450 ↗

1950
Calatrava

Reference number: 1589
Case: 18-karat red gold, push-down case back, leather strap, Ø 36 mm
Movement: Caliber 12-400, rhodium-plated, côtes de Genève, Seal of Geneva, 18 jewels, regulated in 8 positions, manual winding
Remarks: this watch was a state gift
Estimated value: $9,100 ↗

1938
Hooded Lugs

Case: 18-karat red gold/stainless steel, push-down case back, leather strap, Ø 33 mm
Movement: rhodium-plated, côtes de Genève, manual winding
Remarks: rare two-tone men's watch with hooded lugs
Estimated value: $13,000 ↗

1948
Men's Watch

Reference number: 1569
Case: 18-karat pink gold, push-down case back, leather strap, Ø 36 mm
Movement: Caliber 12-120, rhodium-plated, fausses côtes decoration, glucydur screw balance, 18 jewels, regulated in 8 positions, manual winding
Remarks: very fine, elegant Geneva men's watch
Estimated value: $11,300 →

1946
Chronograph

Reference number: 533
Case: 18-karat rose gold, push-down case back, leather strap, Ø 33 mm
Movement: Caliber 13''', rhodium-plated, fausses côtes decoration, very finely polished and beveled steel chronograph parts, column-wheel control, glucydur screw balance, 23 jewels, regulated in 8 positions, manual winding
Remarks: very fine, extremely rare chronograph with 30-minute counter
Estimated value: $45,500 ➔

1945
Chronograph

Reference number: 130
Case: stainless steel, push-down case back, leather strap, Ø 34 mm
Movement: Caliber 13, rhodium-plated, côtes de Genève, column-wheel control of chronograph, 23 jewels, regulated in 8 positions, manual winding
Remarks: extremely rare chronograph in stainless steel with 30-minute counter; only produced in small quantities; offered with excerpt from Patek Philippe archives
Estimated value: $156,000 ➚

1949
World Time

Reference number: 1415 HU
Case: 18-karat yellow gold, push-down case back, rotating knurled bezel, leather strap, Ø 31 mm
Movement: Caliber 12-200, rhodium-plated, côtes de Genève, 18 jewels, regulated in 8 positions, manual winding
Remarks: wristwatch with world time display; reference cities located on rotating bezel, which can be coordinated with the 24-hour ring
Estimated value: $84,500 ➚

1936
One-Button Chronograph

Reference number: 130
Case: 18-karat yellow gold, tripartite, push-down case back, leather strap, Ø 33 mm
Movement: fausses côtes, column-wheel chronograph control
Remarks: 30-minute counter, tachymeter scale; rare sector dial; 1st chrono with reference number; offered with original gold buckle and excerpt from master registry
Estimated value: $260,000 ➔

1938
Chronograph

Case: 18-karat yellow gold, tripartite, push-down case back, leather strap, Ø 33 mm
Movement: Caliber 13-130, rhodium-plated, fausses côtes decoration, column-wheel control of chronograph, finely finished, beveled steel chronograph components, mirror-polished screws, 23 jewels, regulated in 6 positions
Remarks: important chronograph with 30-minute counter and tachymeter scale; 2-tone dial; sold to Astrua in Torino, Italy; offered in original box with certificate
Estimated value: $78,000 ➔

1939
Split-Seconds Chronograph

Reference number: 1436
Case: 18-karat yellow gold, tripartite, push-down case back, leather strap, Ø 33 mm
Movement: rhodium-plated, fausses côtes, column-wheel chronograph control, Seal of Geneva, 25 jewels, regulated in 8 positions
Remarks: extremely rare with 30-minute counter and split-seconds; only 38 made in yellow gold with silver dial, 13 with applied Breguet numerals; offered with 18-karat gold buckle
Estimated value: $325,000 ➔

1942
Chronograph Perpetual Calendar

Reference number: 1518
Case: stainless steel, push-down case back, leather strap, Ø 35 mm
Movement: rhodium-plated, côtes de Genève, column-wheel control of chronograph, 23 jewels, regulated in 8 positions, manual winding
Remarks: extremely rare chronograph with 30-minute counter, perpetual calendar, and moon phase; only 2 ever produced in stainless steel; most expensive wristwatch auctioned in Germany
Estimated value: $2,600,000 ➚

1945
Chronograph Perpetual Calendar

Reference number: 1518
Case: 18-karat yellow gold, push-down case back, leather strap, Ø 35 mm
Movement: rhodium-plated, côtes de Genève, column-wheel control of chronograph, 23 jewels, regulated in 8 positions, manual winding
Remarks: extremely rare chronograph with 30-minute counter, perpetual calendar, and moon phase; from the first series of a total of only 281 ever made
Estimated value: $390,000 ➔

1940
Chronograph

Reference number: 591
Case: 18-karat yellow gold, push-down case back, leather strap, Ø 34 mm
Movement: Caliber 13, rhodium-plated, côtes de Genève, column-wheel control of chronograph, 23 jewels, regulated in 8 positions, manual winding
Remarks: chronograph with 30-minute counter and tachymeter scale; this watch was offered with certificate, leather strap, and gold buckle
Estimated value: $117,000 →

1947
Perpetual Calendar

Reference number: 1526
Case: 18-karat yellow gold, tripartite, push-down case back, leather strap
Movement: Caliber 12-120 QP, rhodium-plated, fausses côtes decorations, mirror-polished screws, 18 jewels, regulated in 8 positions
Remarks: first serially manufactured Patek Philippe perpetual calendar/moon phase; btwn. 1941 and 1952 only 210 pieces of this reference produced; offered with excerpt from master registry
Estimated value: $156,000 →

1946
Men's Watch

Reference number: 1461
Case: 18-karat yellow gold, push-down case back, leather strap, Ø 32 mm
Movement: Caliber 10-200, rhodium-plated, côtes de Genève, manual winding
Remarks: fine men's watch with subsidiary seconds and teardrop-shaped strap lugs
Estimated value: $5,850 ↗

1950
Men's Watch

Reference number: 1578
Case: 18-karat rose gold, push-down case back, leather strap, Ø 35 mm
Movement: Caliber 12-120, rhodium-plated, côtes de Genève, 18 jewels, regulated in 8 positions, manual winding
Remarks: elegant men's watch with extremely rare black dial; this watch was offered with an excerpt from the Patek Philippe master registry
Estimated value: $9,750 ↗

1950
Men's Watch

Reference number: 1584
Case: 18-karat yellow gold, push-down case back, leather strap, Ø 34 mm
Movement: Caliber 12-120, rhodium-plated, côtes de Genève, 18 jewels, manual winding
Remarks: elegant men's watch; this watch was offered with Patek Philippe master registry excerpt and gold clasp
Estimated value: $8,450 ↗

1947
Men's Watch

Reference number: 1569
Case: 18-karat yellow gold, push-down case back, leather strap, Ø 36 mm
Movement: Caliber 12-120, rhodium-plated, côtes de Genève, 18 jewels, regulated in 8 positions, manual winding
Remarks: elegant men's watch with Patek Philippe gold buckle; this watch was offered with an original certificate
Estimated value: $8,450 →

1922
Officer's Watch

Case: 18-karat yellow gold, push-down case back, leather strap, Ø 32 mm
Movement: Caliber 12''', gold plating, frosted finish, wolf's tooth winding gears, indirect sweep seconds, compensation screw balance, manual winding
Remarks: extremely rare early men's wristwatch, Officer model
Estimated value: $28,600 →

1949
Snail

Reference number: 1491
Case: 18-karat yellow gold, tripartite, push-down case back, leather strap, Ø 34 mm
Movement: Caliber 27SC, rhodium-plated, fausses côtes decoration, mirror-polished screws, Seal of Geneva
Remarks: elegant men's watch with sweep seconds; this watch was delivered to Somazzi and also has that signature on the dial
Estimated value: $9,100 ↗

1951
Snail

Reference number: 1491
Case: 18-karat red gold, push-down case back, leather strap, Ø 34 mm
Movement: Caliber 12-120, rhodium-plated, côtes de Genève, 18 jewels, regulated in 8 positions, manual winding
Remarks: red gold men's watch with curved strap lugs; this watch was offered with a Patek Philippe gold buckle and a Patek Philippe certificate
Estimated value: $9,100 ↗

1952
Men's Watch

Reference number: 2459
Case: 18-karat yellow gold, push-down case back, leather strap, Ø 36 mm
Movement: Caliber 12-120, rhodium-plated, côtes de Genève, 18 jewels, manual winding
Remarks: fine men's watch with subsidiary seconds
Estimated value: $8,450 ↗

1960
Men's Watch

Reference number: 2568
Case: 18-karat yellow gold, push-down case back, leather strap, Ø 33 mm
Movement: Caliber 10-200, rhodium-plated, côtes de Genève, 18 jewels, regulated in 8 positions, manual winding
Remarks: elegant men's watch
Estimated value: $4,550 ↗

1949
Chronometer

Reference number: 2452
Case: 18-karat red gold, push-down case back, leather strap, Ø 35 mm
Movement: Caliber 10-200, rhodium-plated, côtes de Genève, manual winding
Remarks: nearly unworn red gold men's watch; this watch was offered in its original box and chronometer bulletin
Estimated value: $15,600 ↗

1946
Men's Watch

Reference number: 1536
Case: 18-karat rose gold, push-down case back, leather strap, Ø 34 mm
Movement: Caliber 12-120 SC, rhodium-plated, fausses côtes decoration, indirect sweep seconds, lépine toothed winding wheels, 20 jewels, regulated in 8 positions, manual winding
Remarks: with indirect sweep seconds after Elysée Piguet; originally sold January 30, 1943; offered with Patek Philippe buckle and excerpt from master registry
Estimated value: $11,700 →

1954
Men's Watch

Reference number: 2515
Case: 18-karat yellow gold, tripartite, push-down case back, leather strap, Ø 36 mm
Movement: Caliber 12-400, rhodium-plated, fausses côtes decoration, polished screws
Remarks: rare men's watch with subsidiary seconds and blue dial; stepped case
Estimated value: $10,400 →

1945
Chronograph Waterproof

Reference number: 1463
Case: stainless steel, screw-down case back, leather strap, Ø 35 mm
Movement: Caliber 13-130, rhodium-plated, côtes de Genève, column-wheel control of chronograph, Seal of Geneva, 23 jewels, regulated in 8 positions, manual winding
Remarks: extremely rare stainless steel chronograph with 30-minute counter and tachymeter scale
Estimated value: $169,000 →

1953
Waterproof Chronograph

Reference number: 1463
Case: 18-karat rose gold, screw-down steel case back, leather strap, Ø 35 mm
Movement: Caliber 13-130, rhodium-plated, côtes de Genève, column-wheel control of chronograph, Seal of Geneva, 23 jewels, regulated in 8 positions, manual winding
Remarks: chronograph with 30-minute counter and tachymeter scale; this watch was offered in its original box with an excerpt from the master registry
Estimated value: $169,000 →

1960
Chronograph Perpetual Calendar

Reference number: 2499
Case: 18-karat yellow gold, push-down case back, leather strap, Ø 35 mm
Movement: Caliber 13-130, rhodium-plated, column-wheel chronograph control, Seal of Geneva, 23 jewels, regulated in 8 positions, manual
Remarks: extremely rare (only 349 made); 30-minute counter, complete calendar, moon phase; offered with elephant hair strap and gold buckle in original box
Estimated value: $520,000 ➔

1960
Chronograph

Reference number: 1579
Case: 18-karat yellow gold, push-down case back, leather strap, Ø 36 mm
Movement: Caliber 13-130, rhodium-plated, côtes de Genève, column-wheel control of chronograph, manual winding
Remarks: elegant chronograph with 30-minute counter; this watch was offered in its original box
Estimated value: $78,000 ➚

1965
Calatrava

Reference number: 570
Case: 18-karat white gold, push-down case back, leather strap, Ø 35 mm
Movement: Caliber 27SC, rhodium-plated, côtes de Genève, Seal of Geneva, 18 jewels, regulated in 8 positions, manual winding
Remarks: very fine Calatrava in white gold
Estimated value: $26,000 ➚

1960
Calatrava

Reference number: 2509
Case: stainless steel, bipartite, screw-down case back, leather strap, Ø 35 mm
Movement: Caliber 12-400, rhodium-plated, fausses côtes decoration, mirror-polished screws, Seal of Geneva, 18 jewels, regulated in 8 positions
Remarks: important men's watch; additional movement protection cap; this watch was offered with an excerpt from the Patek Philippe master registry
Estimated value: $39,000 ➚

1960
Amagnetic

Reference number: 3417
Case: stainless steel, screw-down case back, leather strap, Ø 35 mm
Movement: Caliber 12-400, rhodium-plated, côtes de Genève, gold escapement, Seal of Geneva, 18 jewels, regulated in 8 positions, manual winding
Remarks: rare anti-magnetic men's watch; soft-iron core to protect the movement from magnetic fields; this watch was offered with a Patek Philippe buckle
Estimated value: $26,000 ➚

1954
Men's Automatic Watch

Reference number: 2526
Case: platinum, screw-down case back, leather strap, Ø 35 mm
Movement: Caliber 12-600 AT, rhodium-plated, côtes de Genève, Seal of Geneva, 18-karat gold rotor, 30 jewels, regulated in 8 positions, automatic winding
Remarks: extremely rare automatic watch with diamond-set dial; only 600 pieces of this watch were manufactured, 20 of which were in platinum
Estimated value: $104,000 ➚

1957
Men's Automatic Watch

Reference number: 2526
Case: 18-karat red gold, screw-down case back, leather strap, Ø 35 mm
Movement: Caliber 12-600 AT, rhodium-plated, côtes de Genève, gold rotor, 30 jewels, regulated in 8 positions, automatic winding
Remarks: extremely rare men's watch with subsidiary seconds; only 30 of this watch were manufactured; enamel dial
Estimated value: $52,000 ➔

1965
Men's Watch

Reference number: 3429
Case: platinum, bipartite, screw-down case back, leather strap, Ø 35 mm
Movement: Caliber 27-460, rhodium-plated, fausses côtes decoration, mirror-polished screws, 18-karat gold rotor, Seal of Geneva, 37 jewels, regulated in 8 positions
Remarks: important men's watch with subsidiary seconds in platinum case
Estimated value: $39,000 ➚

1963
Men's Watch

Reference number: 3419
Case: stainless steel, comprising several parts, screw-down case back, leather strap, Ø 34 mm
Movement: Caliber 27AM-400, rhodium-plated, fausses côtes, Seal of Geneva
Remarks: specially shaped case with quick-change system for band protected by Swiss patent no. 340786; this watch was offered with a leather strap and a stainless steel Milanaise bracelet
Estimated value: $5,200 ↗

1960
Men's Automatic Watch

Reference number: 3444
Case: 18-karat yellow gold, screw-down case back, leather strap, Ø 35 mm
Movement: Caliber 27-460, rhodium-plated, côtes de Genève, automatic winding
Remarks: automatic men's watch; 18-karat gold rotor
Estimated value: $9,100 ↗

1952
Men's Automatic Watch

Reference number: 3514
Case: 18-karat yellow gold, screw-down case back, leather strap, Ø 34 mm
Movement: Caliber 27-460M, rhodium-plated, côtes de Genève, gold rotor, Seal of Geneva, 37 jewels, regulated in 8 positions, automatic winding
Remarks: elegant men's watch with subsidiary seconds and date at 3 o'clock
Estimated value: $10,400 ↗

1958
Men's Automatic Watch

Reference number: 2584
Case: 18-karat rose gold, push-down case back, gold Milanaise bracelet, Ø 36 mm
Movement: Caliber 12-600AT, rhodium-plated, côtes de Genève, gold rotor, 30 jewels, regulated in 8 positions, automatic winding
Remarks: elegant men's watch in rose gold with subsidiary seconds
Estimated value: $19,500 ↗

1941
Men's Automatic Watch

Reference number: 3425
Case: 18-karat red gold, screw-down case back, leather strap, Ø 33 mm
Movement: Caliber 27-460, rhodium-plated, côtes de Genève, gold rotor, automatic winding
Remarks: extremely rare men's watch, only about 100 pieces manufactured in red gold; this watch was offered with a Patek Philippe folding clasp and a Patek Philippe certificate
Estimated value: $16,900 ↗

1956
Men's Automatic Watch

Reference number: 2551
Case: 18-karat yellow gold, screw-down case back, leather strap, Ø 36 mm
Movement: Caliber 12-600AT, rhodium-plated, côtes de Genève, gold rotor, 30 jewels, regulated in 8 positions, automatic winding
Remarks: rare automatic men's watch with gold rotor; this watch was offered with a Patek Philippe certificate and buckle
Estimated value: $14,300 ↗

1962
Men's Automatic Watch

Reference number: 3445
Case: 18-karat yellow gold, screw-down case back, leather strap, Ø 35 mm
Movement: Caliber 27-460M, rhodium-plated, côtes de Genève, gold rotor, Seal of Geneva, 37 jewels, regulated in 8 positions, automatic winding
Remarks: elegant men's watch with subsidiary seconds and date at 3 o'clock
Estimated value: $13,000 →

1955
Automatic

Reference number: 2551
Case: 18-karat white gold, screw-down case back, leather strap, Ø 36 mm
Movement: Caliber 12-600AT, rhodium-plated, fausses côtes decoration, Gyromax balance, 30 jewels, regulated in 8 positions, automatic winding
Remarks: very fine, extremely rare Geneva wristwatch with automatic winding, diamond indexes
Estimated value: $20,800 ↗

1951
Men's Watch

Reference number: 2493
Case: 18-karat yellow gold, push-down case back, leather strap, 31 x 40 mm
Movement: Caliber 10-200, rhodium-plated, côtes de Genève, Seal of Geneva, 18 jewels, regulated in 8 positions, manual winding
Remarks: elegant men's watch in a square case; this watch was offered in its original box
Estimated value: $5,200 ↗

1960
Men's watch

Reference number: 2540/2
Case: 18-karat rose gold, bipartite, push-down case back, leather strap, 31 x 41 mm
Movement: Caliber 27SC, rhodium-plated, fausses côtes decoration, mirror-polished screws, Seal of Geneva, 18 jewels, regulated in 8 positions
Remarks: extremely rare men's watch in rounded square (carré galbé) case; so-called TV watch; this watch was offered with an excerpt from the Patek Philippe master registry
Estimated value: $10,400 ↗

1963
Perpetual Calendar

Reference number: 3448
Case: 18-karat white gold, push-down case back, leather strap, Ø 37 mm
Movement: Caliber 27-460Q, rhodium-plated, côtes de Genève, Seal of Geneva, 37 jewels, regulated in 8 positions, automatic winding
Remarks: extremely rare perpetual calendar and moon phase only manufactured 586 times (1962-1982), fewer in white gold; offered in original box with original certificate
Estimated value: $221,000 →

1979
Sensa Luna

Reference number: 3448
Case: 18-karat white gold, tripartite, Ø 37 mm
Movement: Caliber 27-60Q, rhodium-plated, fausses côtes, 18-karat gold rotor, Seal of Geneva, 37 jewels, regulated in 8 positions
Remarks: extremely rare perpetual calendar with small date at 6; window display weekday and month; only 586 Ref. 3448 produced 1962-1982; only 6 known of version shown without moon phase, 2 in white gold
Estimated value: $221,000 →

1985
Perpetual Calendar with Moon Phase

Reference number: 3450
Case: 18-karat yellow gold, push-down case back, leather strap, Ø 37 mm
Movement: Caliber 27-460Q, rhodium-plated, fausses côtes decoration, Gyromax balance, 37 jewels, regulated in 8 positions, automatic winding
Remarks: very fine, extremely rare Geneva men's watch with a perpetual calendar, leap year display, and moon phase
Estimated value: $169,000 →

1977
Nautilus "Jumbo"

Reference number: 3700/1
Case: stainless steel, screw-down case back, stainless steel bracelet, Ø 42 mm
Movement: Caliber 28-255C, rhodium-plated, fausses côtes decoration, Gyromax balance, 36 jewels, regulated in 8 positions, automatic winding
Remarks: legendary luxury stainless steel watch from the 1970s: uniquely large, uniquely expensive
Estimated value: $32,500 ↗

1975
Quartz Beta 21

Reference number: 3587/2
Case: 18-karat yellow gold, screw-down case back, yellow gold bracelet, 42 x 42 mm
Movement: tuning-fork quartz movement, Caliber Beta 21, stop-second system, 13 jewels, battery-driven
Remarks: Patek Philippe belonged to the development consortium of a Swiss quartz movement Beta 21 and produced some watches with that movement
Estimated value: $19,500 →

1982
Men's Automatic Watch

Reference number: 3569-3
Case: 18-karat white gold, screw-down case back, leather strap, Ø 35 mm
Movement: Caliber 1-350, rhodium-plated, côtes de Genève, gold rotor, Seal of Geneva, 28 jewels, regulated in 8 positions, automatic winding
Remarks: elegant men's watch in white gold with blue dial
Estimated value: $6,500 ↗

1971
Men's Automatic Watch

Reference number: 3445
Case: 18-karat white gold, screw-down case back, leather strap, Ø 35 mm
Movement: Caliber 27-460M/PM, rhodium-plated, gold rotor, Seal of Geneva, 37 jewels, regulated in 8 positions, automatic winding
Remarks: men's watch with date and subsidiary seconds; gold rotor
Estimated value: $16,900 →

1969
Ellipse d'Or

Reference number: 3548/1
Case: 18-karat yellow gold, push-down case back, 18-karat yellow gold bracelet, 32 x 27 mm
Movement: Caliber 23-300MP, rhodium-plated, fausses côtes decoration, Seal of Geneva, mirror-polished screws, shock protection for the escape wheel, 18 jewels, regulated in 8 positions, manual winding
Remarks: gold Ellipse d'Or with gold bracelet
Estimated value: $6,500 →

1976
Men's Watch

Reference number: 3585
Case: 18-karat white gold, screw-down case back, leather strap, Ø 36 mm
Movement: Caliber 1-350, rhodium-plated, côtes de Genève, gold rotor, Seal of Geneva, 28 jewels, regulated in 8 positions, automatic winding
Remarks: elegant men's watch in a white gold case
Estimated value: $6,500 ↗

1973
Men's watch with Caliber Beta 21

Reference number: 3603
Case: 18-karat yellow gold, push-down case back, leather strap, 37 x 33 mm
Movement: Caliber 8167, rhodium-plated, polished, tuning fork quartz movement, 13 jewels
Remarks: extremely rare gold men's watch with early Beta 21 tuning fork caliber with date display at 3 o'clock; this watch was offered with a gold Patek Philippe buckle
Estimated value: $9,100 ↗

1993
Perpetual Calendar

Reference number: 5040
Case: 18-karat yellow gold, push-down case back with sapphire crystal, leather strap, 35 x 43 mm
Movement: Caliber 240, rhodium-plated, côtes de Genève, 27 jewels, regulated in 8 positions, automatic winding
Remarks: automatic men's watch with perpetual calendar, moon phase, and 24-hour display; small quantity manufactured
Estimated value: $39,000 →

1989
Jump Hour

Case: 18-karat red gold, push-down case back, leather strap, 28 x 38 mm
Movement: rhodium-plated, côtes de Genève, manual winding
Remarks: limited men's watch with jump digital display in a tonneau-shaped case; this watch was offered in its original box with papers
Estimated value: $26,000 →

1993
Chronograph Perpetual Calendar

Reference number: 3970E
Case: 18-karat red gold, screw-down case back, leather strap, Ø 36 mm
Movement: Caliber CH27-70Q, rhodium-plated, Seal of Geneva, 24 jewels, regulated in 8 positions, manual winding
Remarks: rare chronograph with 30-minute counter, 24-hour display, complete calendar, moon phase, leap year; offered with gold buckle in original box with certificate
Estimated value: $130,000 →

1995
Perpetual Calendar Retrograde

Reference number: 5050
Case: 18-karat white gold, sapphire crystal case back, leather strap, Ø 35 mm
Movement: Caliber 315/136, rhodium-plated, côtes de Genève, gold rotor, Seal of Geneva, 31 jewels, regulated in 8 positions, automatic
Remarks: men's watch with perpetual calendar with retrograde date display; window displays of weekday, month, and leap year; moon phase at 6 o'clock; small quantity manufactured
Estimated value: $78,000 ↗

PIAGET

Georges Piaget started making pocket watches in the solitary environment of a farmhouse in La Côte-aux-Fées in 1874. By the beginning of the twentieth century, twenty-six years later, the company had also begun making wristwatches. Its success, however, remained modest.
That was to change in the 1950s, when Piaget began producing its own calibers under the leadership of Gerald and Valentin Piaget. The company's profits ascended steeply.
In 1957, Piaget introduced its own manually wound movement. Caliber 9P was very flat and thus followed the fashion of the era. Such flat watch movements were to become Piaget's signature element in the ensuing years. The same was true of automatic caliber 12P, introduced in 1959. To this day, it remains one of the thinnest automatic calibers in existence. Its full height of only 2.3 millimeters was only possible because Piaget's watchmakers chose to use a microrotor. Its design was complicated, however: in order to guarantee a secure hold for the rotor despite the low height of the movement, the watchmakers used bearings on both sides. So that the small rotor had enough weight to wind the mainspring, they crafted it in 24-karat gold. Thus they achieved a new automatic movement only 0.2 millimeters higher than manually wound Caliber 9P.
Piaget's business continued to develop so positively that the brothers purchased Baume & Mercier in 1964. The family wanted to continue making progress in the area of flat automatic movements, however, and because competitor Lassalle was manufacturing an even thinner movement, in 1983 Piaget simply bought this company's complete stock of discontinued production. Super-thin Caliber 2000—despite being outfitted with a central rotor—was even flatter than the Piaget Caliber 12P at 2.08 millimeters.
Piaget's watches can be basically divided into thin, elegant men's models and exceptional jeweled timepieces. Robust sports watches, on the other hand, have never been something one can expect to find in this company's collection.
In 1988, some of Piaget's stock was sold to the Vendôme Group (today called Richemont), to which Cartier also belonged. Then in 1993, Piaget was completely sold to this group. Ironically, just a few years previously, the firm would have been able to buy the ailing Cartier group itself—something that would have made Piaget the concern's flagship instead of only a member and movement supplier as it is today.

Founded:
1874 by Georges-Édouard Piaget in La Côte-aux-Fées

Headquarters:
La Côte-aux-Fées, today Geneva (Plan-les-Ouates)

Status:
active without interruption (Richemont Luxury Group)

Specialty:
extra- and ultra-thin watches, jewelry watches, extra-thin automatic watches

Significant inventions:
- 1957 Caliber 9P
- 1960 Caliber 12P, automatic movement with microrotor, height 2.3 mm

1947
Automatic Calendar

Case: stainless steel, screw-down case back, leather strap, Ø 35 mm
Movement: Felsa Caliber 694, rhodium-plated, 25 jewels, automatic winding
Remarks: rare early automatic watch with manual date adjustment by pushing the button at 8 o'clock
Estimated value: $520 →

1955
Men's Watch

Case: 18-karat yellow gold, push-down case back, leather strap, Ø 34 mm
Movement: ETA Caliber 2390, rhodium-plated, manual winding
Remarks: gold men's watch with portrait of Saudi king Ibn Saud on the dial; the coat of arms of the royal house is engraved on the case back
Estimated value: $1,820 →

1957
Men's Watch

Case: 18-karat white gold, push-down case back, leather strap
Movement: Piaget Caliber 9P, ultra flat (2 mm), côtes de Genève, 18 jewels, regulated in 5 positions
Remarks: very fine men's watch with fitted, domed case; recessed crown
Estimated value: $2,210 →

archive photo

1967
Men's Watch

Case: 18-karat white gold, push-down case back, leather strap
Movement: Piaget Caliber 9P, ultra flat (2 mm), côtes de Genève, 18 jewels, regulated in 5 positions
Remarks: very fine extra-flat men's watch with clous de Paris guilloché decoration
Estimated value: $3,250 ↗

archive photo

1960
Men's Watch

Case: 18-karat white gold, push-down case back, leather strap
Movement: Piaget Caliber 9P, ultra flat (2 mm), côtes de Genève, 18 jewels, regulated in 5 positions
Remarks: very fine extra-flat men's watch with clous de Paris guilloché decoration on the bezel
Estimated value: $2,600 ↗

archive photo

1997
Coin Watch

Case: 20-dollar 1907 gold coin, push-down case back, leather strap, Ø 35 mm
Movement: Piaget Caliber 9P, rhodium-plated, fausses côtes decoration, glucydur balance, 18 jewels, regulated in 6 positions, manual winding
Remarks: ultraflat specialty in a hollowed out coin
Estimated value: $3,250 →

1963
Men's Watch Protocole

Case: 18-karat yellow gold, push-down case back, leather strap
Movement: Piaget Caliber 9P, ultra flat (2 mm), côtes de Genève, 18 jewels, regulated in 5 positions
Remarks: very fine extra-flat men's watch with guilloché case and dial
Estimated value: $2,080 →

archive photo

1963
Men's Watch

Case: 18-karat yellow gold, push-down case back, covered strap lugs, leather strap
Movement: Piaget Caliber 9P, ultra flat (2 mm), côtes de Genève, 18 jewels, regulated in 5 positions
Remarks: very fine extra flat men's watch
Estimated value: $3,250 →

archive photo

PIERCE

In 1888, Léon Lévy and his brothers founded the Manufacture des Montres & Chronographes Pierce SA in Biel. The Lévy brothers' decision to dedicate themselves completely to making chronographs was courageous and forward-looking at the same time: chronographs were still relatively unimportant at that point. Their heyday began only with mass motorization and the introduction of auto racing.

Pierce's chronographs, which were at first outfitted with purchased movements, quickly achieved high recognition. Over time, more and more companies began offering chronographs in their collections, so supply in the 1920s overtook demand. These watches became harder to sell, and a price war broke out. In order to stabilize the prices, watchmaking's governing bodies fought back.

The result was that Pierce, classified as an *établisseur,* was no longer supplied with *ébauches:* the company had no other choice but to develop and manufacture its own movements in the early 1930s.

Well-known and popular among collectors are the 13-line Pierce chronograph calibers 130 and 134. These autonomous movements are interesting from a technical point of view since they are outfitted with a vertical clutch, allowing the movement to get along without a separate coupling wheel.

Yet Pierce was not limited to only chronographs. The creativity of the company's technicians was evident in other developments as well. In the 1940s, Pierce introduced its own automatic movement, Caliber 861. Although pendulum and rotor windings had both already technically matured, and the preferred technology of the time—Pierce made its own "shaking" automatic according to a patent from 1933. Here, the winding weight was led straight along two racks. In one direction, a small toothed rack wound the mainspring.

Another special development was the Correctomatic from the 1950s, which may have looked like a chronograph, but was something quite different. The little buttons at 2 and 4 o'clock were used for rate regulation instead. By pushing one of the two buttons, the index was moved. One of the buttons made the movement run faster, and the other slowed it down. This was controlled by a little mechanism that put pressure on the index, allowing for fine adjustment of the watch.

Another specialty was the Duofon: an alarm wristwatch that let its wearer adjust the alarm volume. The wearer could choose whether the Duofon should ring loudly or more discreetly, in which case it only rattled quietly. A small window at 6 o'clock displayed whether the watch was set to loud or quiet.

In the 1990s, Pierce was still introducing various models outfitted with different calibers made by other manufacturers. Among these was a small series of automatic alarm wristwatches outfitted with Omega/Lemania Caliber 980. Like the Omega Memomatic, Pierce's timepiece could also be set precisely to the minute.

Founded:
1888 by Léon Lévy and Brothers in Biel/Bienne

Headquarters:
Biel/Bienne

Status:
revived (Léon Lévy & Frères)

Specialty:
chronographs, flyer and service watches, automatics, wristwatch alarms

Significant inventions:
- 1931 chronograph with a vertical clutch
- 1933 linear automatic
- 1956 Duofon wristwatch alarm
- 1959 Correctomatic

Leading models:
Correctomatic, Duofon

1940
Chronograph

Case: 18-karat red gold, push-down case back, leather strap, Ø 37 mm
Movement: nickel-plated, côtes de Genève, column-wheel control of chronograph, manual winding
Remarks: gold chronograph with 60-minute counter
Estimated value: $2,340 ➔

1940
Chronograph

Case: gold-plated, stainless steel push-down case back, leather strap, Ø 37 mm
Movement: rhodium-plated, decorated, column-wheel control of chronograph, manual winding
Remarks: chronograph with rare 60-minute counter
Estimated value: $1,080 ➔

1945
Chronograph

Case: chrome-plated, stainless steel push-down case back, leather strap, Ø 36 mm
Movement: nickel-plated, column-wheel control of chronograph, manual winding
Remarks: chronograph with rare 60-minute counter
Estimated value: $520 ➔

1940
Chronograph

Case: chrome-plated, stainless steel push-down case back, leather strap, Ø 34 mm
Movement: nickel-plated, côtes de Genève, column-wheel control of chronograph, manual winding
Remarks: unusual one-button chronograph with button at 4 o'clock
Estimated value: $520 ➔

1950
Correctomatic

Case: gold-plated, stainless steel screw-down case back, leather strap, Ø 35 mm
Movement: nickel-plated, with regulating lever system via button, manual winding
Remarks: on the Correctomatic the watch's rate can be regulated forward or backward using a button on the case; Pierce's own regulating lever system makes this possible
Estimated value: $1,300 ➔

1965
Duofon

Case: stainless steel, push-down case back, leather strap, Ø 33 mm
Movement: Caliber 135, nickel-plated, twin spring barrels, gong, manual winding
Remarks: alarm wristwatch with Pierce manufacture caliber; movement has integrated gong; second crown allows the alarm volume to be chosen
Estimated value: $1,560 ➔

1946
Automatic

Reference number: 311199
Case: chrome-plated, stainless steel screw-down case back, leather strap, Ø 37.5 mm
Movement: Caliber 861, nickel-plated, 17 jewels, automatic winding
Remarks: rare early automatic watch by Pierce; Caliber 861 has a so-called "shaking" weight that tensions the mainspring via a toothed rack
Estimated value: $650 ➔

1976
Alarm

Case: stainless steel, screw-down case back, leather strap
Movement: Caliber AS 5008, 17 jewels, automatic winding
Remarks: automatic wristwatch alarm with AS movement
Estimated value: $455 ➔

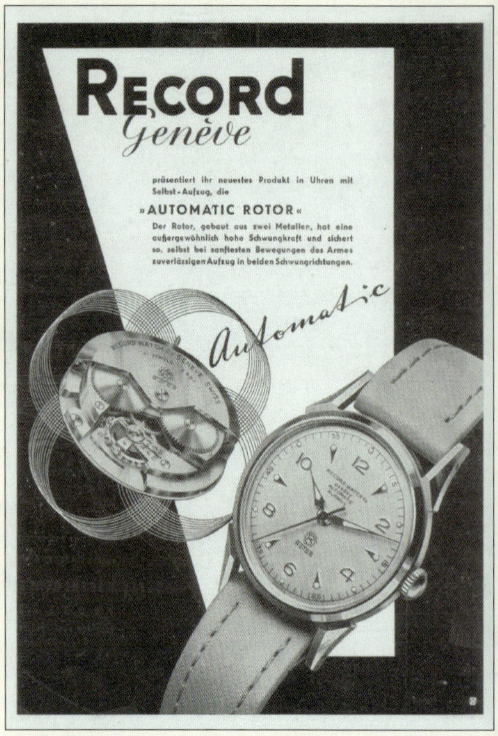

RECORD

Record's watches are only familiar to collectors these days. This company was founded in 1903 as Record Dreadnought Watch, a somewhat martial moniker. As the name suggests, these watches had no need to shy away from their competitors: the highest rate precision possible was the topmost goal of the small watch manufacturer. In the view of those in charge, this could only be achieved if the components were manufactured as precisely as possible. Manufacturing tolerances that were too great were, according to the company, one of the main elements negatively influencing a watch's precision.

This concept resulted in the desired success in the 1920s. At the time, Record Watch already maintained three production facilities manufacturing both pocket and wristwatches. Many timepieces by Record were delivered with an official chronometer certificate. But even without one, Record's watches still achieved the best rates, and this was certainly a reason why Record Watch provided various military outfits all over the world with service watches.

Alongside chronometers, Record is above all known for its calendar watches, the so-called Datofix models. Chronographs, among them some with split-seconds functions, were also something Record had in its program, utilizing mainly manually wound movements by Venus. Both three-handed watches and chronographs were outfitted with complete calendar mechanisms and an optional moon phase.

However, most of the movements came from the company's own production. Record's first movement with automatic winding was Caliber 161/162, presented in 1944, and then successor model 171/172. This was an automatic movement with winding according to the Harwood principle, using a weight on a central bearing that was not able to swing freely, but rather oscillated between two spring-loaded limiting bumpers. This hammering movement is the reason that the system is also called a hammer automatic.

Beginning in 1952 the successor to the hammer automatics was Caliber 174, a modern rotor automatic form that was also available with date and power reserve displays.

In 1961, Record Watch Co. lost its independence and was sold to Longines in Saint-Imier. After that, watches were produced under the double signature "Record Longines." These timepieces were no longer outfitted with *manufacture* calibers, but with mainly ETA technology. For Longines, this formerly reputable brand was only a subsidiary that produced reasonably priced watches, and it was later completely abandoned.

Founded:
1903 in Tramelan

Headquarters:
Tramelan

Status:
dissolved

Specialty:
chronometers, calendar watches, flyer and service watches, automatic watches

Leading models:
Datofix

1945
Datofix
Case: 18-karat red gold, push-down case back, leather strap, Ø 35 mm
Movement: Caliber 1066, nickel-plated, côtes de Genève, jewels set in chatons, manual winding
Remarks: men's watch with complete calendar and moon phase
Estimated value: $2,470 ➔

1950
Datofix
Case: stainless steel, screw-down case back, leather strap, Ø 34 mm
Movement: Caliber 1076, nickel-plated, manual winding
Remarks: men's watch with complete calendar and moon phase
Estimated value: $1,950 ➔

1945
Datofix
Case: 18-karat red gold, push-down case back, leather strap, Ø 35 mm
Movement: nickel-plated, manual winding
Remarks: red gold men's watch with complete calendar and moon phase
Estimated value: $1,820 ➔

1945
Chronograph with Complete Calendar
Case: stainless steel, gold-plated, push-down case back, leather strap, Ø 35 mm
Movement: Caliber 88, nickel-plated, côtes de Genève, column-wheel control of chronograph, manual winding
Remarks: elegant chronograph with 30-minute and 12-hour counters; complete calendar and moon phase
Estimated value: $2,340 ➔

1955
Bidynator Automatic
Case: 18-karat yellow gold, screw-down case back, leather strap, Ø 37 mm
Movement: Felsa caliber, nickel-plated, automatic winding
Remarks: automatic men's watch with Felsa Bidynator automatic caliber; complete calendar and moon phase
Estimated value: $2,080 ➔

1948
General Service Watch W.W.W.
Case: stainless steel, screw-down case back, textile strap, Ø 35 mm
Movement: Caliber 022K, rhodium-plated, manual winding
Remarks: men's watch of the British armed forces "W.W.W."
Estimated value: $520 ➔

1950
Chronograph
Case: 18-karat yellow gold, push-down case back, leather strap, Ø 38 mm
Movement: nickel-plated, côtes de Genève, column-wheel control of chronograph, manual winding
Remarks: heavy gold chronograph with 45-minute counter
Estimated value: $2,340 ➔

1950
Rattrapante Chronograph
Case: stainless steel, push-down case back, leather strap, Ø 37 mm
Movement: Caliber 185, rhodium-plated, finely polished steel chronograph parts, column wheel control, manually wound
Remarks: rare, fine Geneva chronograph with rattrapante function
Estimated value: $8,840 ➚

ROLEX

Rolex cannot be compared to any other company in the watch industry. This brand has long said good-bye to any of the usual standards, and its products are always preceded by a legendary reputation. Rolex stands for glamour and wealth, but also for precision and innovation. And there is one man to thank for all of that: Hans Wilsdorf.

This self-made man—at the age of twelve he was orphaned and forced to find his way alone in the world—first learned everything about timekeepers when working at a large watch wholesaling outfit in La Chaux-de-Fonds. With this experience under his belt, the German later went to London to build up a watch wholesale business with a rich partner named Davis. He soon put all his money on wristwatches, for they were in demand among gentlemen after the Boer War in Africa had shown them to be so practical—and they had a strong masculine image.

Wilsdorf focused on small 11-line calibers that he purchased from the Swiss producer Aegler in Biel. They were even superior to many pocket watches in rate precision and robustness. The proof: Wilsdorf had one of his Aegler Caliber 1910s tested as a chronometer at a Swiss observatory, and it became the first wristwatch caliber to receive a chronometer certificate. In 1914, he repeated this experiment in Kew, England. This watch also did excellently and received a Class A certificate.

Wilsdorf had successfully introduced and maintained the Rolex brand name and the logo with the five-pointed crown for eleven years when he started having customs trouble in the British Empire in 1919 and had to move his business to Switzerland. In Geneva, he founded Montres Rolex SA. Five years previously, Wilsdorf had won Aegler over as his exclusive movement supplier. He made Rolex Watch Co., Aegler SA out of this firm.

Rolex's success is based on two great inventions that helped wristwatches to their breakthrough and made them what they are today: water-resistant and self-winding. Wilsdorf developed the first absolutely water-resistant Oyster case to house the Aegler movements. Because the constant winding of the movement threatened this water resistance, Wilsdorf and his technicians developed the first serially mature rotor automatic in 1931—and the Oyster Perpetual was born. From this point on, it would be the standard, and to this day most automatic watches function according to its principle.

The Oyster remains the company's most famous model, especially in the form of diver's watches and chronographs. The early Prince and Cellini models were not only reliable, they were also elegant timepieces.

Thanks to consistent development, the Oyster remained the measuring stick for robust sports watches. Modified as the Submariner, it became the first professional diver's watch. As the Oyster Sea Dweller it even received a helium valve and special gas mixtures for professional diving.

Rolex watches have also impressively proven their reliability on many expeditions. They have already been on top of Mount Everest, at both Poles, and at the deepest point of the earth, the Mariana Trench more than 11,000 meters down. They have even been in space.

Wilsdorf eventually donated his shares of the company to the Wilsdorf Foundation he had established since he had no heirs. After his death in 1960, André J. Heiniger took the reins. In 1992, his son Patrick Heiniger followed him as president.

Founded:
- 1905 as Wilsdorf & Davis, London
- 1908 Rolex brand trademarked
- 1919 company renamed Montres Rolex SA

Headquarters:
Geneva

Status:
active without interruption

Specialty:
chronometers, pocket watches, GMT watches

Significant inventions:
- 1926 water-resistant Oyster case
- 1931 "perpetual" winding rotor
- 1967 automatic helium valve

Leading models:
Submariner GMT, Explorer, Cosmograph Daytona

1935
Prince Chronometer
Case: 18-karat yellow gold, push-down case back, leather strap, 20 x 40 mm
Movement: rhodium-plated, regulated in 8 positions, manual winding
Remarks: elegant Prince Classic
Estimated value: $4,550 ➔

1936
Prince Railway Observatory Quality
Reference number: 1527M
Case: 18-karat white/yellow gold, push-down case back, leather strap, 22 x 42 mm
Movement: rhodium-plated, 15 jewels, regulated in 6 positions, manual winding
Remarks: rare Prince Railway in perfect quality with observatory-quality movement
Estimated value: $13,000 ➔

1950
Prince Aerodynamic Chronometer
Reference number: 3361
Case: 18-karat rose gold, bipartite, push-down case back, leather strap, 19 x 46 mm
Movement: Caliber 310, rhodium-plated, Ultra Prima quality, 18 jewels, regulated in 6 positions
Remarks: asymmetric; introduced in 1939; Caliber 310 produced 1932-38, also housed in later Ultra Prima quality Prince models; offered with 18-kt Rolex buckle
Estimated value: $9,100 ➔

1935
Prince Brancard Chronometer Extraprima Observatory Quality
Reference number: 971U
Case: silver/red gold, push-down case back, leather strap, 26 x 43 mm
Movement: rhodium-plated, 15 jewels, regulated in 6 positions, manual winding
Remarks: elegant men's watch in Brancard case; observatory-quality movement, regulated in six positions
Estimated value: $13,000 ➚

1930
Prince 1/4 Century Club
Reference number: 3937
Case: 14-karat yellow gold, push-down case back, leather strap, 22 x 44 mm
Movement: rhodium-plated, regulated in 6 positions, manual winding
Remarks: rare Rolex Prince as 1/4 Century Club version commemorating the company's 25 years as part of the Eaton Concern
Estimated value: $5,915 ➔

1930
Prince Railway Jump Hours Observatory Quality
Reference number: 1587HS
Case: 18-karat red/white gold, push-down case back, leather strap, 23 x 42 mm
Movement: rhodium-plated, regulated in 6 positions, manual winding
Remarks: rare men's watch in a stepped rectangular case with jump, digital hours and subsidiary seconds
Estimated value: $20,800 ➔

1934
Prince Brancard Chronometer Extraprima
Reference number: 971
Case: 18-karat yellow gold, push-down case back, leather strap, 25 x 43 mm
Movement: rhodium-plated, regulated in 8 positions, manual winding
Remarks: rare Prince chronometer in Art Deco style
Estimated value: $10,400 ➔

1945
Prince Brancard Chronometer Jump Hours Extra Prima Observatory Quality
Reference number: 1491
Case: 9-kt gold, push-down back, 25 x 43 mm
Movement: rhodium-plated, regulated in 6 positions, manual winding
Remarks: rare, with digital jump hours; offered in original box with original Biel rate certificate
Estimated value: $39,000 ➚

1932
Oyster Perpetual Chronometer For Ronchi, Milan

Reference number: 3347
Case: 18-karat red gold, screw-down case back, leather strap, Ø 29 mm
Movement: rhodium-plated, automatic winding
Remarks: extremely rare early Oyster with double hour markers; III and IX are vertical on bezel but horizontal on dial; rare double signature "Ronchi Milano"
Estimated value: $6,500 ➔

1932
Oyster Perpetual Super Precision

Reference number: 3353
Case: stainless steel/gold, screw-down case back, stainless steel link bracelet, Ø 29 mm
Movement: rhodium-plated, automatic winding
Remarks: extremely rare early automatic Oyster with subsidiary seconds
Estimated value: $5,200 ➔

1940
Oyster Perpetual Chronometer Bubble Back Hooded Lugs

Case: stainless steel/gold, screw-down case back, leather strap, Ø 32 mm
Movement: rhodium-plated, automatic winding
Remarks: Oyster Perpetual with Roman numerals and hooded lugs
Estimated value: $3,900 ➔

1948
Oyster Perpetual Chronometer Bubble Back

Reference number: 3372
Case: stainless steel, screw-down case back, leather strap, Ø 32 mm
Movement: rhodium-plated, automatic winding
Remarks: so-called Oyster Bubble Back
Estimated value: $3,250 ➔

1949
Oyster Perpetual Chronometer with Mickey Mouse Dial

Reference number: 3131
Case: 14-karat rose gold, screw-down case back, leather strap, Ø 32 mm
Movement: rhodium-plated, automatic winding
Remarks: rare Oyster Perpetual with original Mickey Mouse dial; this watch was offered in its original box and with a Rolex buckle
Estimated value: $9,100 ➔

1940
Oyster Perpetual Chronometer Bubble Back

Reference number: 3131
Case: 14-karat red gold, screw-down case back, leather strap, 32 x 39 mm
Movement: rhodium-plated, automatic winding
Remarks: rare red gold Bubble Back
Estimated value: $3,380 ➔

1948
Oyster Perpetual Bubble Back

Reference number: 3133
Case: stainless steel, screw-down case back, gold bezel, stainless steel/gold link bracelet, Ø 32 mm
Movement: rhodium-plated, automatic winding
Remarks: Oyster Bubble Back with legible, large, luminous numerals
Estimated value: $3,900 ➔

1932
Military Antimagnetic Chronograph

Reference number: 2508
Case: stainless steel, tripartite, push-down case back, leather strap, Ø 35 mm
Movement: nickel-plated, finely finished, column-wheel chronograph control, finely finished steel components, polished screws
Remarks: rare chronograph with 30-minute counter; this chronograph is one of the first chronograph models manufactured by Rolex
Estimated value: $39,000 ➔

1935
Chronograph

Reference number: 3371
Case: 18-karat yellow gold, tripartite, push-down case back, leather strap, Ø 35 mm
Movement: nickel-plated, finely finished, column-wheel control of chronograph, finely finished steel chronograph components, polished screws, 17 jewels
Remarks: rare very fine chronograph with 30-minute counter, tachymeter and blue telemeter scales; this watch was offered with a gold Rolex buckle
Estimated value: $19,500 →

1926
Chronograph Antimagnétique

Reference number: 2057
Case: 18-karat yellow gold, tripartite, push-down case back, leather strap, 32 x 37 mm
Movement: nickel-plated, finely finished, column-wheel control of chronograph, finely finished, beveled steel chronograph components, polished screws
Remarks: very rare one-button chronograph with 30-minute counter; chronograph button integrated into crown
Estimated value: $26,000 →

1950
Chronograph Antimagnétique

Case: 18-karat yellow gold, push-down case back, leather strap, Ø 30 mm
Movement: nickel-plated, column-wheel control of chronograph, manual winding
Remarks: extremely rare chronograph with 30-minute counter and tachymeter scale
Estimated value: $15,600 →

1948
Chronograph

Reference number: 4099
Case: stainless steel, push-down case back, leather strap, Ø 35 mm
Movement: rhodium-plated, column-wheel control of chronograph, manual winding
Remarks: rare chronograph in stainless steel with exceptional strap lugs
Estimated value: $19,500 →

1934
Chronograph Antimagnetic

Reference number: 2508
Case: 18-karat yellow gold, push-down case back, leather strap, Ø 36 mm
Movement: nickel-plated, column-wheel control of chronograph, manual winding
Remarks: one of the first chronographs manufactured by Rolex with 30-minute counter and tachymeter scale
Estimated value: $18,200 →

1934
Chronograph Antimagnétique

Reference number: 3834
Case: 18-karat yellow gold, push-down case back, leather strap, Ø 32 mm
Movement: rhodium-plated, manual winding
Remarks: extremely rare chronograph with 30-minute counter and tachymeter scale
Estimated value: $15,600 →

1947
Oyster Chronograph

Reference number: 4500
Case: stainless steel, screw-down case back, leather strap, Ø 36 mm
Movement: Caliber R23, rhodium-plated, column-wheel control of chronograph, manual winding
Remarks: rare Oyster chronograph with 30-minute counter and tachymeter scale
Estimated value: $32,500 →

1956
Chronograph

Reference number: 2811
Case: stainless steel, push-down case back, leather strap, Ø 32 mm
Movement: rhodium-plated, finely polished and beveled steel chronograph parts, column-wheel control, screw balance, manual winding
Estimated value: $15,600 →

1948
Oyster Chronograph Antimagnetic

Reference number: 4500
Case: 18-karat yellow gold, screw-down case back, leather strap, Ø 36 mm
Movement: rhodium-plated, finely polished and beveled steel chronograph parts, column-wheel control, screw balance, 17 jewels, manual winding
Estimated value: $32,500 ➔

1949
Oyster Chronograph

Reference number: 4500
Case: stainless steel, screw-down case back, gold bezel, leather strap, Ø 36 mm
Movement: rhodium-plated, column-wheel control of chronograph, manual winding
Remarks: Oyster chronograph with 30-minute counter; this watch was offered in its original box
Estimated value: $18,200 ➔

1940
Oyster Chronograph Anti-Magnétique

Reference number: 3668
Case: stainless steel, screw-down case back, red gold index bezel, screw-in crown, Ø 35 mm
Movement: rhodium-plated, column-wheel chronograph control, beveled steel components, polished screws, 17 jewels
Remarks: with 30-minute counter; telemeter scale and screw-in Oyster Plus crown; offered in a contemporary box
Estimated value: $22,100 ➔

1949
Chronograph Dato Compax

Reference number: 4768
Case: stainless steel, tripartite, push-down case back, leather strap, Ø 35 mm
Movement: Caliber 72C, rhodium-plated, column-wheel chronograph control, beveled steel components, 17 jewels
Remarks: rare, with 30-minute, 12-hour counters; complete calendar with date hand and window for weekday, month; only 220 produced
Estimated value: $52,000 ➔

1950
Chronograph with Complete Calendar

Reference number: 4768
Case: 18-karat rose gold, push-down case back, leather strap, Ø 35 mm
Movement: Valjoux Caliber 72C, rhodium-plated, column-wheel control of chronograph, manual winding
Remarks: rare rose gold chronograph with 30-minute and 12-hour counters; complete calendar and tachymeter scale
Estimated value: $45,500 ➔

1950
Perpetual Chronometer Precision

Reference number: 8171
Case: stainless steel, screw-down case back, leather strap, Ø 38 mm
Movement: rhodium-plated, automatic winding
Remarks: extremely rare Oyster with complete calendar, subsidiary seconds, and moon phase; this model was only manufactured in a small series and is difficult to find; this watch has the number 279
Estimated value: $130,000 ➔

1949
Perpetual Chronometer

Reference number: 8171
Case: 18-karat red gold, push-down case back, leather strap, Ø 38 mm
Movement: rhodium-plated, automatic winding
Remarks: extremely rare men's watch manufactured in small series; this watch bears the number 202 and is nearly impossible to find in red gold; this watch was offered in its original box
Estimated value: $91,000 ➔

1952
Oyster Perpetual Datejust Chronometer

Reference number: 6105
Case: 18-karat yellow gold, screw-down case back, leather strap, Ø 35 mm
Movement: rhodium-plated, automatic winding
Remarks: elegant Oyster Datejust in gold; this watch was offered with an original testing certificate from the Biel observatory
Estimated value: $5,200 ➔

1955

Oyster Perpetual Datejust Superlative Chronometer Serpico y Laino

Reference number: 6605
Case: 18-karat yellow gold, screw-down case back, leather strap, Ø 36 mm
Movement: Caliber 1065, rhodium-plated, 25 jewels, regulated in 6 positions, automatic winding
Remarks: Oyster with extremely rare printing on dial
Estimated value: $5,850 ➔

1960

Oyster Perpetual Tru-Date Officially Certified Chronometer

Reference number: 6534
Case: stainless steel, screw-down case back, leather strap, Ø 34 mm
Movement: Caliber 1065, rhodium-plated, 25 jewels, regulated in 6 positions, automatic winding
Remarks: Oyster Tru-Date with rare printing on dial; watch delivered with double signature in conjunction with Tiffany & Co.
Estimated value: $3,250 ➔

1956

Oyster Perpetual Tru-Beat Superlative Officially Certified Chronometer

Reference number: 6556
Case: 18-karat yellow gold, screw-down case back, leather strap, Ø 34 mm
Movement: Caliber 1040, rhodium-plated, 26 jewels, regulated in 6 positions, automatic
Remarks: extremely rare Tru-Beat with jumping sweep seconds, characteristic of the most sought-after models; only a few units in gold
Estimated value: $13,000 ➔

1954

Oyster Perpetual Explorer

Reference number: 6298
Case: stainless steel, screw-down case back, gold bezel, leather strap, Ø 35 mm
Movement: rhodium-plated, automatic winding
Remarks: extremely rare early Explorer with additional printed words on dial "Rotor Self-Winding"
Estimated value: $3,900 ➔

1954

Oyster Perpetual Chronometer Bubble Back Star Dial

Reference number: 6098
Case: 18-karat yellow gold, bipartite, screw-down case back, leather strap, Ø 35 mm
Movement: red gold-plated, winding rotor
Remarks: big Bubble Back in an unusual combination of case with honeycomb dial; extremely rare applied gold hour markers; offered with gold 18-kt Rolex buckle in original box
Estimated value: $18,200 ➔

1955

Datejust Chronometer

Reference number: 6305/1
Case: stainless steel, screw-down case back, stainless steel link bracelet, Ø 36 mm
Movement: Caliber 745, rhodium-plated, automatic winding
Remarks: rare Datejust with guilloché dial; black/red date; white gold bezel; this model was manufactured only from 1953 to 1955
Estimated value: $3,250 ➔

1972

Oyster Perpetual Day-Date Chronometer

Reference number: 1803
Case: 18-karat yellow gold, screw-down case back, leather strap, Ø 35 mm
Movement: Caliber 1556, rhodium-plated decorated, 26 jewels, regulated in 6 positions, automatic winding
Remarks: Oyster Perpetual Day-Date in gold
Estimated value: $4,550 ➔

1961

Oyster Perpetual Day-Date Chronometer

Reference number: 1803
Case: 18-karat yellow gold, screw-down case back, leather strap, Ø 36 mm
Movement: Caliber 1555, rhodium-plated, decorated, 26 jewels, regulated in 6 positions, automatic winding
Remarks: men's watch with display of day and date
Estimated value: $6,500 ➔

1956
Oyster Perpetual Tru-Beat Superlative Chronometer "By Official Test"

Reference number: 6556
Case: stainless steel, screw-down case back, stainless steel link bracelet, Ø 35 mm
Movement: Caliber 1040, rhodium-plated, 26 jewels, regulated in 6 positions, automatic
Remarks: extremely rare Oyster with jumping sweep seconds; the dial is printed with the rare and very unusual words "by official test"
Estimated value: $9,100 ➔

1946
Presidential Seal

Reference number: 6085
Case: 14-karat yellow gold, bipartite, screw-down case back, leather strap, Ø 33 mm
Movement: nickel-plated, decorated, polished screws, winding rotor
Remarks: important Oyster with enameled dial of the seal of the United States in cloisonné technique by C. Poluzzi; due to the great deal of work done by hand, this watch is considered unique
Estimated value: $130,000 ➔

1972
Oyster Precision

Reference number: 6426
Case: stainless steel, screw-down case back, leather strap, Ø 35 mm
Movement: Caliber 1225, rhodium-plated, manual winding
Remarks: rare Precision with dragon motif on dial
Estimated value: $5,200 ➚

1956
Date Chronometer

Reference number: 1503
Case: 18-karat yellow gold, screw-down case back, Milanaise bracelet, Ø 34 mm
Movement: Caliber 1560, rhodium-plated, decorated, 26 jewels, regulated in 6 positions, automatic winding
Remarks: Rolex Date with double signature: Rolex and Serpico y Laino
Estimated value: $5,200 ➔

1953
Oyster Perpetual Chronometer

Reference number: 6084
Case: 18-karat yellow gold, bipartite, screw-down case back, 18-karat yellow gold link bracelet
Movement: nickel-plated, decorated, polished screws, winding rotor
Remarks: gold Bubble Back with original Oyster gold link bracelet and index bezel; this watch was offered in its original box
Estimated value: $7,150 ➔

1953
Oyster Perpetual Chronometer Officially Certified

Reference number: 6062
Case: stainless steel, screw-down case back, stainless steel link bracelet, Ø 36 mm
Movement: rhodium-plated, automatic winding
Remarks: extremely rare Oyster with complete calendar and moon phase; only 350 pieces of this model were produced in red gold and only a few in stainless steel between 1950 and 1953
Estimated value: $156,000 ➔

1953
Oyster Perpetual Chronometer

Reference number: 6062
Case: 18-karat yellow gold, bipartite, screw-down case back, leather strap, Ø 35 mm
Movement: rhodium-plated, polished screws, winding rotor
Remarks: extremely rare men's watch with complete calendar; date hand; windows for weekday, month; moon phase at 6; only 350 of this model produced in red gold between 1950 and 1953
Estimated value: $130,000 ➚

1954
Oyster Chronograph Antimagnetic Jean-Claude Killy

Reference number: 6036
Case: stainless steel, screw-down case back, link bracelet, Ø 36 mm
Movement: Caliber 72C, rhodium-plated, column-wheel chronograph control, manual
Remarks: rare chronograph with 30-minute and 12-hour counters; complete calendar with date, weekday, and month; model named for ski-racing great of the same name
Estimated value: $195,000 ➔

1953

Oyster Chronograph

Reference number: 6236
Case: 18-karat yellow gold, tripartite, screw-down case back, leather strap, Ø 36 mm
Movement: Caliber 72C, rhodium-plated, column-wheel control of chronograph
Remarks: extremely rare chronograph with 30-minute and 12-hour counters; complete calendar with date hand and window for weekday, month; the so-called Jean-Claudy Killy; only 170 produced in yellow gold
Estimated value: $208,000 ➔

1955

Chronograph Antimagnetic

Reference number: 3335
Case: stainless steel, screw-down case back, stainless steel link bracelet, Ø 37 mm
Movement: Caliber Valjoux 22, rhodium-plated, finely polished and beveled steel chronograph parts, column-wheel control, screw balance, 17 jewels, manual winding
Remarks: extremely rare, very fine men's watch with chronograph
Estimated value: $78,000 ➚

1965

Chronograph (Pre-Daytona)

Case: 14-karat yellow gold, screw-down case back, Ø 36 mm
Movement: rhodium-plated, column-wheel control of chronograph, manual winding
Remarks: blue tachymeter scale
Estimated value: $39,000 ➔

1966

Chronograph (Pre-Daytona)

Reference number: 6238
Case: stainless steel, screw-down case back, stainless steel link bracelet, Ø 36 mm
Movement: Caliber 722.1, rhodium-plated, column-wheel control of chronograph, 17 jewels, regulated in 3 positions, manual winding
Remarks: chronograph with 30-minute and 12 hour counters
Estimated value: $26,000 ➔

1962

Oyster Chronograph Antimagnetic

Reference number: 6234
Case: stainless steel, tripartite, screw-down case back, link bracelet, Ø 36 mm
Movement: Valjoux Caliber 72, nickel-plated, finely finished, column-wheel chronograph control, 17 jewels
Remarks: rare chronograph with 30-minute and 12-hour counters; tachymeter and telemeter scales in Oyster case
Estimated value: $26,000 ➔

1965

Chronograph (Pre-Daytona)

Reference number: 6238
Case: 14-karat yellow gold, screw-down case back, gold link bracelet, Ø 32 mm
Movement: Caliber 72B, rhodium-plated, column-wheel control of chronograph, manual winding
Remarks: chronograph with 30-minute and 12-hour counters
Estimated value: $78,000 ➔

1979

Oyster Cosmograph Daytona

Reference number: 6265/6263
Case: stainless steel, screw-down case back, stainless steel link bracelet, Ø 37 mm
Movement: Caliber 727, rhodium-plated, column-wheel control of chronograph, manual winding
Remarks: Daytona chronograph with 30-minute and 12-hour counters; this watch was offered in its original box with a certificate, operation manual, and original invoice
Estimated value: $26,000 ➚

1979

Oyster Cosmograph Daytona

Reference number: 6263
Case: stainless steel, screw-down case back, stainless steel link bracelet, Ø 37 mm
Movement: Caliber 727, rhodium-plated, finely polished and beveled steel chronograph parts, column-wheel control, glucydur screw balance, manual winding
Estimated value: $26,000 ➔

1980
Oyster Cosmograph Daytona

Reference number: 6265
Case: stainless steel, screw-down case back, stainless steel link bracelet, Ø 37 mm
Movement: Caliber 727 rhodium-plated, finely polished and beveled steel chronograph parts, column-wheel control, glucydur screw balance, manual winding
Remarks: the so-called "Paul Newman" version with a special dial design is considered one of the most coveted Daytona models
Estimated value: $104,000 →

1960
Oyster Cosmograph Daytona Paul Newman

Reference number: 6240
Case: stainless steel, screw-down case back, stainless steel link bracelet, Ø 37 mm
Movement: rhodium-plated, column-wheel control of chronograph, manual winding
Remarks: rare Daytona chronograph with black tachymeter scale on bezel; this watch was offered with service chit and confirmation of authenticity
Estimated value: $78,000 ↗

1977
Oyster Super Chronometer Cosmograph Daytona

Reference number: 6263
Case: 18-karat yellow gold, tripartite, screw-down case back, yellow gold link bracelet, Ø 38 mm
Movement: Caliber 727/1531, rhodium-plated, column-wheel chronograph control, 17 jewels, regulated in 3 positions
Remarks: rare, with 30-minute, 12-hour counters; tachymeter scale on bezel; screw-in crown, buttons; offered in original box with certificate
Estimated value: $58,500 →

1982
Oyster Chronometer Cosmograph

Reference number: 6263/6265
Case: 18-karat yellow gold, screw-down case back, gold link bracelet, Ø 37 mm
Movement: Caliber 727, rhodium-plated, column-wheel control of chronograph, manual winding
Remarks: rare Cosmograph with 30-minute and 12-hour counters; tachymeter scale on bezel
Estimated value: $52,000 ↗

1957
Oyster Chronometer Cosmograph

Reference number: 6241
Case: 18-karat yellow gold, screw-down case back, leather strap, Ø 37 mm
Movement: Caliber 727, rhodium-plated, column-wheel control of chronograph, manual winding
Remarks: rare Cosmograph with 30-minute and 12-hour counters; tachymeter scale on bezel
Estimated value: $39,000 ↗

1980
Cosmograph

Reference number: 6265
Case: 18-karat yellow gold, screw-down case back, leather strap, Ø 37 mm
Movement: Caliber 722, rhodium-plated, column-wheel control of chronograph, manual winding
Remarks: rare Cosmograph with 30-minute and 12-hour counters; tachymeter scale on bezel
Estimated value: $45,500 ↗

1953
Oyster Perpetual Precision Explorer

Reference number: 6150
Case: stainless steel, screw-down case back, leather strap, Ø 35 mm
Movement: rhodium-plated, automatic winding
Remarks: early Explorer model
Estimated value: $7,800 →

1958
Oyster Perpetual Explorer Super Precision

Reference number: 5500
Case: stainless steel, screw-down case back, stainless steel link bracelet, Ø 34 mm
Movement: Caliber 1530, rhodium-plated, automatic winding
Remarks: rare Explorer
Estimated value: $5,200 →

1953
Oyster Perpetual Chronometer Explorer
Reference number: 6350
Case: stainless steel, screw-down case back, leather strap, Ø 35 mm
Movement: rhodium-plated, automatic winding
Remarks: rare early Explorer model with three-dimensional structure on dial
Estimated value: $9,100 ➔

1954
Oyster Perpetual Turn-O-Graph
Reference number: 6202
Case: stainless steel, tripartite, screw-down case back, rotating bezel, leather strap
Movement: nickel-plated, finely finished, polished screws, winding rotor
Remarks: early Oyster Turn-O-Graph with rotating black bezel with markers
Estimated value: $5,200 ➔

1977
Oyster Perpetual Datejust Chronometer Thunderbird
Reference number: 1625
Case: stainless steel, screw-down case back, 18-karat gold bezel, Ø 36 mm
Movement: rhodium-plated, automatic winding
Remarks: rare Oyster Thunderbird with black dial
Estimated value: $3,900 ➔

1972
Oyster Perpetual Datejust Chronometer Thunderbird
Reference number: 1625
Case: 18-karat yellow gold, screw-down case back, leather strap, Ø 36 mm
Movement: Caliber 1570, rhodium-plated, automatic winding
Remarks: early Thunderbird in gold
Estimated value: $5,070 ➔

1958
Oyster Perpetual Milgauss Superlative Officially Certified Chronometer
Reference number: 6541
Case: stainless steel, screw-down back, rotating bezel, link bracelet, Ø 36 mm
Movement: Caliber 1066M, rhodium-plated, 25 jewels, regulated in 6 positions, automatic winding
Remarks: extremely rare first-edition Milgauss without foudroyante hand
Estimated value: $52,000 ➔

1966
Oyster Perpetual Milgauss
Reference number: 1019
Case: stainless steel, screw-down case back, stainless steel link bracelet, Ø 37 mm
Movement: rhodium-plated, automatic winding
Remarks: rare early Milgauss; the movement is protected by a soft iron core and thus protected from magnetic fields to 1,000 Gauss
Estimated value: $23,400 ➔

1953
Oyster Perpetual Turn-O-Graph
Reference number: 6202
Case: stainless steel, screw-down case back, stainless steel link bracelet, Ø 36 mm
Movement: Caliber A296, rhodium-plated, automatic winding
Remarks: rare Turn-O-Graph with Rolex riveted steel bracelet
Estimated value: $7,800 ➔

1955
Oyster Perpetual Submariner James Bond
Reference number: 6205
Case: stainless steel, screw-down case back, stainless steel link bracelet, Ø 36 mm
Movement: Caliber A260, rhodium-plated, automatic winding
Remarks: this watch got its nickname by appearing in several early James Bond films
Estimated value: $26,000 ➔

1957
Oyster Perpetual Submariner 100 m/330 ft

Reference number: 6536/1
Case: stainless steel, screw-down case back, rotating bezel, stainless steel link bracelet, Ø 37 mm
Movement: Caliber 1030, rhodium-plated, 25 jewels, regulated in 6 positions, automatic winding
Remarks: early chronometer version of the Submariner
Estimated value: $14,300 →

1961
Oyster Perpetual Submariner 100 m/330 ft

Reference number: 5508
Case: stainless steel, screw-down case back, stainless steel link bracelet, Ø 37 mm
Movement: Caliber 1530M, rhodium-plated, 25 jewels, regulated in 6 positions, automatic winding
Remarks: unique Submariner with Explorer dial and remarkable engraving on case back: "Casma Lima Peru 4/2/63"
Estimated value: $28,600 ↗

1964
Oyster Perpetual Submariner 200 m = 660 ft Explorer Dial

Reference number: 5513
Case: stainless steel, screw-down case back, stainless steel link bracelet, Ø 39 mm
Movement: Caliber 1520, rhodium-plated, 26 jewels, regulated in 6 positions, automatic winding
Remarks: Submariner with extremely rare Explorer-style dial variation
Estimated value: $26,000 →

1977
Oyster Perpetual Submariner 660ft = 200m

Reference number: 1680
Case: stainless steel, screw-down case back, rotating bezel, stainless steel link bracelet, Ø 39 mm
Movement: Caliber 1570, rhodium-plated, fausses côtes decoration, 26 jewels, regulated in 6 positions, automatic winding
Estimated value: $7,800 ↗

1959
Oyster Perpetual Submariner

Reference number: 6538
Case: stainless steel, screw-down case back, rotating bezel, Ø 38 mm
Movement: Caliber 1030, rhodium-plated, 25 jewels, regulated in 6 positions, automatic winding
Remarks: unique Submariner with original, nearly unknown dial variation; so-called James Bond appeared in several early Bond films
Estimated value: $104,000 →

1978
Sea Dweller Submariner 2000

Reference number: 1665
Case: stainless steel, screw-down case back, rotating bezel, Ø 39 mm
Movement: Caliber 1665, rhodium-plated, automatic winding
Remarks: rare early Sea Dweller diver's watch with red printing on the dial; this watch was offered with a Rolex maintenance chit
Estimated value: $19,500 →

1975
Oyster Perpetual Submariner 660ft = 200m

Reference number: 5513
Case: stainless steel, screw-down case back, rotating bezel, Velcro tape, Ø 39 mm
Movement: Caliber 1520, rhodium-plated, polished, glucydur balance, automatic winding
Remarks: extremely rare military diving watch of the British Special Boat Service
Estimated value: $45,500 ↗

1979
Oyster Perpetual Submariner Date

Reference number: 1680
Case: 18-karat yellow gold, screw-down case back, rotating bezel, Velcro tape, Ø 39 mm
Movement: Caliber 1570, rhodium-plated, fausses côtes decoration, 26 jewels, adjusted in 8 positions, automatic winding
Estimated value: $19,500 →

1970

Oyster Perpetual Date Explorer II

Reference number: 1655
Case: stainless steel, screw-down case back, rotating bezel, Oyster steel bracelet (Ref. 78360), Ø 39 mm
Movement: Caliber 1570, rhodium-plated, finely finished, polished screws, 26 jewels, regulated in 8 positions, automatic winding
Remarks: rare Explorer II, so-called Steve McQueen with Twinlock crown and noticeable hand to indicate the second time zone
Estimated value: $19,500 ➔

1958

Oyster Perpetual GMT Master PanAm Officially Certified Chronometer

Reference number: 6542
Case: stainless steel, screw-down case, rotating bezel, link bracelet, Ø 38 mm
Movement: Caliber 1035, rhodium-plated, 25 jewels, regulated in 6 positions, automatic
Remarks: extremely rare GMT-Master produced as a special series for PanAm; fewer than 200 were made for this special series
Estimated value: $52,000 ➔

1966

Oyster Perpetual GMT-Master

Reference number: 1675
Case: 18-karat yellow gold, screw-down case back, rotating bezel, gold link bracelet, Ø 39 mm
Movement: rhodium-plated, automatic winding
Remarks: gold GMT-Master with 24-hour display; a second time zone can be set by the rotating bezel
Estimated value: $65,000 ➔

1985

Oyster Perpetual Date GMT-Master

Reference number: 16758
Case: 18-karat yellow gold, screw-down case back, rotating bezel, yellow gold link bracelet, Ø 39 mm
Movement: Caliber 3075, rhodium-plated, 27 jewels, adjusted in 7 positions, automatic winding
Estimated value: $10,400 ➚

1956

Oyster Perpetual GMT-Master Officially Certified Chronometer

Reference number: 6542
Case: stainless steel, screw-down case, rotating bezel, link bracelet, Ø 38 mm
Movement: Caliber 1030, rhodium-plated, 25 jewels, regulated in 6 positions, automatic winding
Remarks: original acrylic rotating bezel; 2nd time zone set via 24-hour hand and bezel
Estimated value: $26,000 ➔

1960

Oyster Perpetual GMT-Master

Reference number: 1675
Case: stainless steel, screw-down case back, rotating bezel, stainless steel link bracelet, Ø 39 mm
Movement: rhodium-plated, automatic winding
Remarks: a second time zone can be set using the 24-hour hand and the rotating, two-tone bezel; the colors blue and red guarantee quick differentiation of day and night
Estimated value: $5,850 ➔

1983

Oyster Perpetual Date Explorer II

Reference number: 1655
Case: stainless steel, screw-down case back, rotating bezel, Oyster steel bracelet (Ref. 78360), Ø 39 mm
Movement: Caliber 1570, rhodium-plated, finely finished, polished screws, 26 jewels, regulated in 8 positions, automatic winding
Remarks: rare Explorer II, so-called Steve McQueen with Twinlock crown and noticeable hand to indicate the second time zone
Estimated value: $18,200 ➔

1984

Explorer II

Reference number: 16550
Case: stainless steel, tripartite, screw-down case back, stainless steel link bracelet, Ø 39 mm
Movement: Caliber 3085, rhodium-plated, finely finished, mirror-polished screws, winding rotor, 27 jewels, regulated in 6 positions
Remarks: Explorer II with rare cream-colored dial; this watch was sold with a guarantee
Estimated value: $7,800 ➔

STOWA

Stowa is among the very few German watch manufacturers that can look back on an unbroken company history. Founded in 1927 by Walter Storz in Hornberg, it was moved to Pforzheim in 1935. He created the brand name Stowa by putting the opening letters of the names Storz and Walter together.

In 1938, the company had grown so much that its owner had an edifice built especially for it. At this time, Stowa already had various wristwatches in its collection, among them a model dominated by a Bauhaus-style dial reflecting the fashion of the time. This dial featured very clean hour numeral typography, which lives on today in the Antea model.

During World War II, Stowa supplied the German air force with pilot's watches and produced marine observation watches. On February 23, 1945, the company's factory was destroyed during the bombing of Pforzheim, which caused Storz to build a new watch production facility in Rheinfelden close to Basel, Switzerland. On November 5 of the same year, Stowa was already able to begin manufacturing watches again in the new premises with the entire workforce that Storz had brought with him from Pforzheim.

Rheinfelden was at first considered a temporary emergency solution. However, after the Pforzheim headquarters were rebuilt, the factory in Rheinfelden became Rheinfelder Uhrteilefabrik (Rufa GmbH), where among other components shock protection subgroups were manufactured for PUW and Durowe calibers. The factory in Rheinfelden was enlarged in 1955 and 1966, and Storz became a founding member of what was to become the German watch industry's governing body.

Stowa offered a wide palette of wristwatches of varying types. It was the robust sports and diver's watches above all that added to the company's good reputation. In 1960, the son of the founder, Werner Storz, took over the company and managed it until 1996.

In the 1970s, the collection comprised about 1,000 different models. In 1974, company founder Walter Storz passed away. The same year Stowa joined the German watch co-op Pallas, which would greatly contribute to the Stowa brand surviving the so-called quartz crisis relatively unscathed.

When Werner Storz retired from the company, Jörg Schauer bought the brand and moved its headquarters to Engelsbrand, a suburb of Pforzheim. Under his leadership, the company began to produce mechanical wristwatches once again with the historical models as their inspiration. The early production even included the partial use of the original printing templates for the dials. Today, Stowa watches are distributed directly via the Internet and at the Engelsbrand headquarters, which is now housed in a brand-new factory building.

5

Founded:
1927 by Walter Storz in Hornberg

Headquarters:
Engelsbrand

Status:
active without interruption (today Jörg Schauer)

Specialty:
flyer and marine observation watches, sports watches, diving watches, jewelry watches

Leading models:
Pilot's Watch Seatime

1940
Pilot's Watch

Case: grey painted brass, tripartite, compressed case back with inner B-Watch engraving, extra long leather strap with stitching, Ø 33 mm
Movement: Unitas Caliber 2812, nickel-plated, côtes de Genève decoration, indirectly powered sweep seconds, swan-neck fine adjustment, polished screws, Breguet hairspring, 20 jewels, gold STOWA engraving
Remarks: pilot's watch from WW II with blued steel hands; Stowa was one of 5 companies supplying the German Air Force
Estimated value: $4,550 ➔

1945
Pilot's Watch

Case: stainless steel, matte, push-down case back, leather strap, Ø 55 mm
Movement: nickel-plated, côtes de Genève, indirect sweep seconds, compensation screw balance, 20 jewels, manual winding
Remarks: pilot's watch of the German Luftwaffe
Estimated value: $2,860 ➔

1949
Armée Française

Case: chrome-plated, stainless steel screw-down case back, leather strap, Ø 33 mm
Movement: Durowe Caliber 422, nickel-plated, finely finished beveled bridges, polished screws, 17 jewels
Remarks: this watch was delivered to the French ground troops as reparation; edition of 2,500 watches; watches no. 1 to 750 were delivered with a 15-jewel movement; watches 751 to 2,500 with a 17-jewel movement; all watches except no. 1 were outfitted with a black dial
Estimated value: $325 ➔

1969
Chrono

Case: stainless steel, bipartite, screw-down case back, textile strap, Ø 39 mm
Movement: Valjoux Caliber 7730, nickel-plated, coulisse-style chronograph, 17 jewels
Remarks: water-resistant chronograph with 30-minute counter; this watch was designed as a simple diver's chronograph using a rotating bezel marked with minute divisions and well visible minute hand; water-resistant to 100 meters
Estimated value: $1,300 ➔

1953
Gold Watch

Case: 14-karat yellow gold, bipartite, push-down case back, leather strap, Ø 35 mm
Movement: nickel-plated, polished screws, indirect sweep seconds, screw balance, 17 jewels
Remarks: gold men's watch housed in an elaborate case with teardrop-shaped, fluted strap lugs
Estimated value: $1,040 ➔

1958
World Time

Case: gold-plated, bipartite, stainless steel screw-down case back, leather strap, Ø 32 mm
Movement: nickel-plated, finely finished beveled bridges, polished screws, 17 jewels
Remarks: rare world time watch with inner 24-hour scale; the respective time zones are displayed in conjunction with the map of the world found in the center of the dial
Estimated value: $520 ➔

1971
Memo Time

Case: chrome-plated, bipartite, stainless steel screw-down case back, leather strap, Ø 36 mm
Movement: Durowe Caliber INT 471-4, nickel-plated, matte finish, 17 jewels
Remarks: simple stopwatch with 60-minute counter shown in window at 12 o'clock; the chronograph function starts when the permanently rotating minute disk is reset via the button at 2 o'clock; the minute counter is coated with luminous substance
Estimated value: $455 ➔

1972
Seatime

Case: chrome-plated, bipartite, screw-down case back, leather strap, Ø 37 mm
Movement: Caliber P 75A, nickel-plated, winding rotor, 21 jewels
Remarks: sports watch in diver's watch look with two-tone rotating bezel; the markers display the hours as well as the minutes in five-minute intervals
Estimated value: $390 ➔

TISSOT

Watch collectors—especially those who are interested in progressive technology—just cannot escape Tissot. This brand, which today belongs to the Swatch Group, made a name for itself again and again in the past with technical innovation. In the 1930s, Tissot was already offering an antimagnetic wristwatch. And in 1944, the Swiss company introduced a movement that many of its competitors oriented their own technology on: Caliber 28.5 was conceived as a module movement, so that without any great effort it would later be able to accommodate additional functions like date or weekday displays. The so-called Autolub caliber, used in the Astrolon watch, also belonged to the technical avant-garde in watchmaking.

Tissot was founded in 1853 by Charles-Félicien Tissot and his son Charles-Emile in Le Locle. Unlike most family histories, in this case it was the son who showed the initiative of founding his own *manufacture.* Charles-Emile lived with his uncle in New York for many years after finishing his watchmaker apprenticeship, during which time he gained a great deal of experience with the most diverse types of watches. The untiring Charles-Emile practiced what we might call active distribution today: his many trips soon made Tissot well-known in countries spanning the globe. Tissot watches were especially popular in Russia—Charles-Emile's son Charles lived in Moscow for many years and built up a subsidiary there. Special regiment watches were created for Russian officers, the backs of which were engraved with the coat of arms of each of the commanders. The October Revolution wiped out Charles's work in one fell swoop, and Tissot's important Russian market broke down.

In 1883, Charles took over the management of Tissot, a company that was now based on quality and progress: up to the end of the nineteenth century, Tissot had won several gold medals and first prizes in watchmaking. In 1917, the Swiss firm launched its first wristwatch, and in 1920 it became a *manufacture*, beginning with large series production of wristwatches. In 1930, it merged with Omega and founded the SSIH (Société Suisse de l'Industrie Horlogère). Within this relationship, Omega received the role of premium brand while Tissot had to satisfy itself with that of the younger sibling, covering the mid-priced segment. In 1932, the cooperative company bought movement maker Lemania, after which Tissot also purchased chronograph calibers from Lemania.

How creative Tissot still was in the area of movement technology was something the *manufacture* proved with the development of the above-mentioned Autolub caliber in the 1970s. Autolub stood for "self-lubricating" and was an important milestone: Tissot had developed a movement made of plastic that did not need to be oiled. This is a conceptual advantage, for old, hardened oil has been proven disadvantageous to a watch's rate. This movement did not immediately become a great success, however, as the plastics that existed at that point in time were not robust enough and the cheaper prices of quartz movements represented more than enough competition for mechanical watches.

Tissot used quartz technology in the 1980s for avant-garde-styled watches such as the Rock Watch, whose case was created from natural stone making each of these timepieces unique. Tissot used granite, gneiss, and numerous exotic stones, and later even mother-of-pearl and briar wood, to make watch cases. These watches were chiefly purchased by people interested in fashion.

Today's collector concentrates on the brand's mechanical classics such as the Navigator with world time display, on which a button at 2 o'clock moves the disk showing the world reference cities. The Navigator spawned an entire family of wristwatches, which remained in production for years and included both chronographs and a wristwatch alarm.

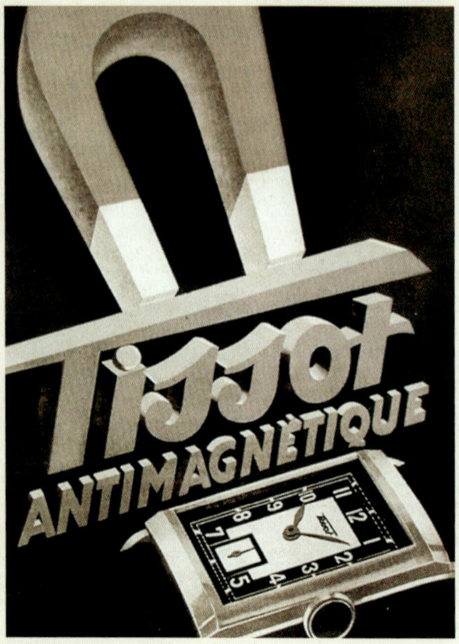

Founded:
1853 by Charles-Félicien and Charles-Emile Tissot in Le Locle

Headquarters:
Le Locle

Status:
active without interruption (Swatch Group)

Specialty:
chronographs, diving watches, robust sports watches, wristwatch alarms

Significant inventions:
- 1970 Astrolon, self-lubricating plastic movement
- 1971 Sideral case
- 1985 Rock Watch

Leading models:
Navigator

1935
Chronograph

Case: stainless steel, leather strap, Ø 37 mm
Movement: gold-plated, frosted finish, column-wheel control of chronograph, manual winding
Remarks: fine chronograph with enamel dial and movable strap lugs
Estimated value: $1,560 ➔

1940
German Army Service Watch

Case: plated, stainless steel push-down case back, leather strap, Ø 34 mm
Movement: nickel-plated, manual winding
Remarks: service watch of the Germany army, numbered “DH35427”
Estimated value: $650 ➔

1945
Chronograph

Case: stainless steel, push-down case back, leather strap, Ø 37 mm
Movement: nickel-plated, column-wheel control of chronograph, manual winding
Remarks: chronograph with 30-minute counter
Estimated value: $1,300 ➔

1945
Chronograph

Case: 18-karat red gold, push-down case back, leather strap, Ø 35 mm
Movement: Caliber 27-41H, nickel-plated, column-wheel control of chronograph, manual winding
Remarks: red gold chronograph with 30-minute and 12-hour counters
Estimated value: $2,340 ➔

1940
Chronograph

Case: stainless steel, push-down case back, leather strap, Ø 37 mm
Movement: gold-plated, frosted finish, column-wheel control of chronograph, manual winding
Remarks: one-button chronograph with 30-minute counter
Estimated value: $2,210 ➔

1945
Chronograph

Case: stainless steel, push-down case back, leather strap, Ø 32 mm
Movement: nickel-plated, column-wheel control of chronograph, manual winding
Remarks: early chronograph with 30-minute counter
Estimated value: $1,040 ➔

1950
Chronograph

Case: 18-karat yellow gold, push-down case back, leather strap, Ø 35 mm
Movement: Caliber C27-41, nickel-plated, column-wheel control of chronograph, manual winding
Remarks: heavy gold chronograph with 30-minute and 12-hour counters
Estimated value: $1,950 ➔

1945
World Time Watch

Case: 14-karat red gold, push-down case back, leather strap, Ø 36 mm
Movement: Caliber 28.5N-21, red gold-plated, automatic winding
Remarks: gold hammer automatic watch; rotating disk with world reference city names (24-hour indication)
Estimated value: $2,340 ➔

1971
Sonorus T12
Case: stainless steel, push-down case back, stainless steel link bracelet
Movement: Caliber AS 1930, nickel-plated, 17 jewels, manual winding
Remarks: rare alarm diver's watch in a heavy stainless steel case; tripartite case back with second case back to withstand great pressure in great depths; diving time is set by the third crown at 11 o'clock
Estimated value: $1,040 →

1969
Sonorus PR516
Case: stainless steel, screw-down case back, stainless steel link bracelet
Movement: Caliber AS 1930, nickel-plated, 17 jewels, manual winding
Remarks: alarm wristwatch in diver's watch look with rotating bezel
Estimated value: $910 →

1975
Navigator Seastar T12 Automatic
Case: stainless steel, screw-down case back, leather strap, Ø 42 mm
Movement: Caliber 798, gold-plated, automatic winding
Remarks: large men's watch with two-tone 24-hour dial; rotating bezel on the inside (flange) with world time display
Estimated value: $520 →

1970
Seastar T12
Case: stainless steel, push-down case back, leather strap, Ø 42 mm
Movement: Lemania Caliber LWO 1281, nickel-plated, manual winding
Remarks: stately chronograph with 30-minute and 12-hour counters
Estimated value: $650 →

1975
Visodate Seamaster T12
Case: stainless steel, screw-down case back, leather strap, 42 x 50 mm
Movement: Caliber 784-2, rhodium-plated, automatic winding
Remarks: diver's watch with rotating inner bezel (flange)
Estimated value: $520 →

1970
Navigator Automatic Chronograph
Case: stainless steel, screw-down case back, leather strap, Ø 38 mm
Movement: Lemania Caliber LWO 1341, nickel-plated, 17 jewels, automatic winding
Remarks: automatic chronograph with sweep minute and subsidiary 12-hour counters
Estimated value: $520 →

1975
Navigator Automatic Chronograph
Case: stainless steel, screw-down case back, stainless steel link bracelet, Ø 40 mm
Movement: Lemania Caliber LWO 1341, rhodium-plated, 17 jewels, automatic winding
Remarks: chronograph with sweep 60-minute and subsidiary 12-hour counters
Estimated value: $585 →

1975
T12 Chronograph
Case: stainless steel, push-down case back, stainless steel link bracelet, Ø 44 mm
Movement: Lemania Caliber LWO 1873, nickel-plated, 17 jewels, manual winding
Remarks: chronograph with 30-minute and 12-hour counters
Estimated value: $585 →

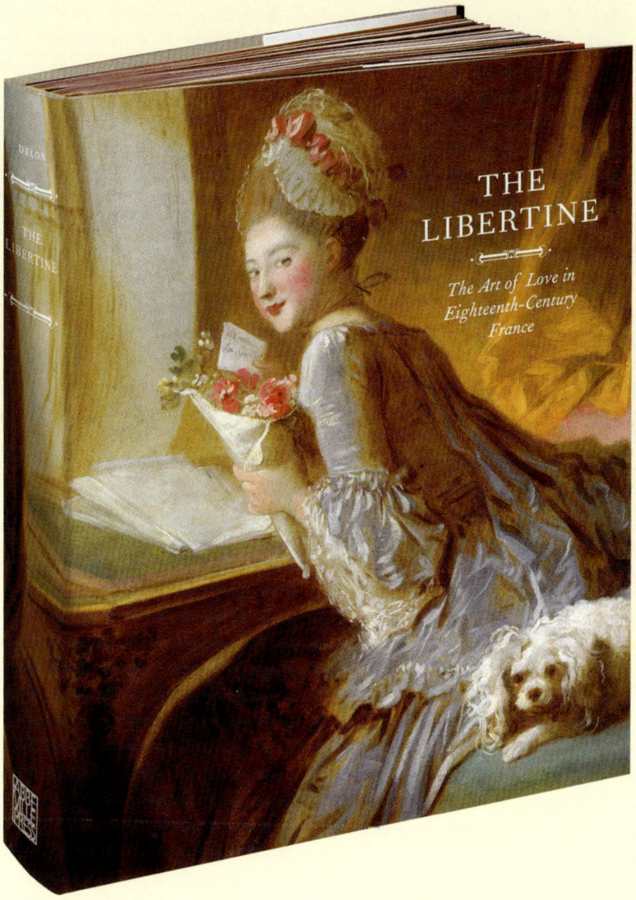

"A delicious fête galante, from a century that equated the feeling of love with an amazing freedom."—*L'Express*

"The game of seduction, the sly glances, the sweet love notes, denuded shoulders, and other pleasures of love are illustrated by works of a refined sensuality by Watteau, Boucher, Greuze, and other poets of the fête galante."—*Le Figaro Magazine*

"The eighteenth century seems to flirt everywhere: in salons, boudoirs, forests; on screens, earthenware, and fans; and the bodies are posed at the reader's disposal. Delectable."—*Le Monde*

"This work is a masterpiece."—*Point De Vue*

The Libertine

The Art of Love in Eighteenth-Century France

A lavishly presented tasting menu of excerpts from eighteenth-century libertine literature, expertly paired with delectable period artworks.

This sumptuous volume re-creates the milieu of the libertine in all its lively decadence, bringing together more than eighty brief selections from eighteenth-century French literature, grouped into eight broad themes—including tales of seduction, fantasies of exotic lands, and the discoveries of youth—and introduced by an eminent French scholar.

Edited by Michel Delon
Foreword by Marilyn Yalom
400 full-color illustrations
496 pages · 9⅝ × 12¼ in.
Cloth · $150.00
ISBN 978-0-7892-1147-7

Published by ABBEVILLE PRESS
137 Varick Street, New York, NY 10013
1-800-ARTBOOK (in U.S. only)
Also available wherever fine books are sold
Visit us at www.abbeville.com

TUDOR

Tudor has always languished in the shadow of its elder sibling Rolex. Regardless, this brand has been able to secure a loyal and enthusiastic following for itself over the decades. It is the hidden qualities of the watches that are so convincing, the special Rolex features that are practically free at Tudor—without the famous crown logo, of course, but also without the surcharge associated with that great moniker. In exchange, the Tudors carry the title of crown prince in their model names: Oyster Prince, Oyster Prince Day-Date, and Oyster Prince Submariner.

Rolex founder Hans Wilsdorf registered many brand names over the course of the decades, but did not end up using all of them. Tudor is the only one remaining active to the present day aside from Rolex. The company's founder had to wait a long time to get his name: when he went to register Tudor, he realized that it had already been given out and applied since 1906 and that there actually already were watches sold under that name. Only in the 1940s did Wilsdorf receive his desired rights to Tudor.

Naturally, it made little sense to purposely undercut the Rolex quality that had been created and maintained for decades for a second brand. And so all Tudor watches literally profit from the technical achievements of the company's leading brand. They don't always have to be the newest and most current developments—after all, with regard to sports watch quality, the Rolex models were always ahead of their time anyway.

The company only "saved" on the movement: instead of a Rolex *manufacture* caliber, Tudor models were mainly outfitted with AS and ETA calibers, which were minutely regulated before being encased.

There is no rule without an exception, and the Tudor brand eventually did get its own movement: in the 1950s, the Tudor Prince Rotor was powered by Tudor Caliber 390. This movement could not hide its origins, which derived from a typical Rolex automatic caliber: various components were very close to Rolex Caliber 1030, which was manufactured at the same time. Parts of the automatic winding technically corresponded to older rotor automatics, though. As Caliber 395, it was outfitted with an additional date display.

In addition to three-hand watches, Tudor also produced chronographs with column-wheel calibers that were very different from Rolex models: Tudor does not use 12-hour counters. Alongside various Oyster models, Tudor also offered an alarm watch in the form of the Advisor model. Such model policies have, however, resulted in Tudor watches being misused for counterfeit Rolex watches.

Founded:
protected brand since 1906; purchased by Hans Wilsdorf in the 1940s

Headquarters:
Geneva

Status:
active without interruption (Rolex SA)

Specialty:
diving watches, sports watches

Leading models:
Advisor, Oyster Prince Submariner, Oyster Date Chrono

1963
Oyster Prince

Case: stainless steel, screw-down case back, stainless steel link bracelet, Ø 35 mm
Movement: nickel-plated, automatic winding
Remarks: sporty men's watch
Estimated value: $1,820 →

1975
Oyster Prince Ranger

Reference number: 9050
Case: stainless steel, screw-down case back, leather strap, Ø 34 mm
Movement: rhodium-plated, automatic winding
Remarks: rare Tudor Ranger with large luminous numerals
Estimated value: $2,600 ↗

1970
Submariner

Case: stainless steel, screw-down case back, rotating bezel, stainless steel link bracelet, Ø 39 mm
Movement: Caliber 2484, nickel-plated, automatic winding
Remarks: diver's watch with rotating bezel
Estimated value: $3,900 →

1965
Oyster Prince Submariner

Reference number: 7928
Case: stainless steel, screw-down case back, rotating bezel, leather strap, Ø 39 mm
Movement: gold-plated, automatic winding
Remarks: heavy diver's watch
Estimated value: $3,900 →

1949
Oyster

Case: 18-karat yellow gold, push-down case back, leather strap, Ø 32 mm
Movement: nickel-plated, 17 jewels, manual winding
Remarks: rare early gold men's watch
Estimated value: $1,300 →

1970
Prince Oyster Date

Reference number: 90814
Case: stainless steel, screw-down case back, rotating bezel, leather strap, Ø 40 mm
Movement: rhodium-plated, decorated, automatic winding
Remarks: large automatic men's watch
Estimated value: $2,340 →

1980
Auto Chrono Time

Case: stainless steel, screw-down case back, leather strap, Ø 40 mm
Movement: Valjoux Caliber 7750, nickel-plated, automatic winding
Remarks: large chronograph with 30-minute and 12-hour counters
Estimated value: $13,000 ↗

1980
Oyster Date Chronograph

Case: stainless steel, screw-down case back, stainless steel link bracelet, Ø 41 mm
Movement: Valjoux Caliber 234, rhodium-plated, column-wheel control of chronograph, manual winding
Remarks: chronograph with 45-minute counter
Estimated value: $9,100 →

ULYSSE NARDIN

At the age of nineteen, the highly gifted watchmaker Ulysse Nardin founded his own company in Le Locle. He had apprenticed with his father and taken the opportunity to train with two highly renowned masters, William Dubois and Louis Richard. It was his father who lent his son the capital to found his own watchmaking enterprise, and he even became an employee of his son later on.

His first watches were delivered to Piguet Frères, who were at the time among the world's best chronometer manufacturers. From 1846 on, Ulysse Nardin made and sold watches under his own name. His repeaters and pocket chronometers were especially popular in South America, Mexico, and the United States. After Ulysse Nardin's death at the age of fifty-three, his only son Paul-David took over the company.

In 1860, Ulysse and Paul-David Nardin had made their first marine chronometer, an object that would become the great specialty of the brand for years to come. The anchor in the company's logo is derived from this maritime association. Marine chronometers by Ulysse Nardin were always among the most precise watches of their time.

Early on, Ulysse Nardin had also begun to manufacture wristwatches. Alongside normal three-handed watches, these were primarily chronographs. To make them, the company used purchased movements, mainly Valjoux and Venus calibers.

After the death of Paul-David in 1920, his sons Alfred, Ernest, and Gaston took over the company's management and turned it into a joint stock company. The Ulysse Nardin brand survived both world wars and the world economic crisis without problem, but the introduction of the quartz marine chronometer in the 1960s broke its back. The beginning of the 1970s saw the end of the more than 100-year-old family tradition. Only the name and the run-down headquarters in Le Locle remained of the once proud company.

These last bits of Ulysse Nardin were purchased in 1983 by Rolf Schnyder, who was able to resuscitate the brand. In a certain way, history repeated itself, for an exceptional watchmaking talent once again heavily contributed to its renewed success: Dr. Ludwig Oechslin. Schnyder discovered him as an apprentice watchmaker in Lucerne. Today, the multitalented Oechslin has completed doctorates in several subjects and is the curator of the prestigious Musée International d'Horlogerie (MIH) in La Chaux-de-Fonds. Rolf Schnyder died in 2011. His wife, Chai Schnyder, was named president of the board of directors, and Patrik Hoffmann was appointed CEO.

Founded:
1846 by Ulysse Nardin in Le Locle

Headquarters:
Le Locle

Status:
active without interruption

Specialty:
marine chronometer, wristwatch chronometer

Significant inventions:
- 1985 Astrolabium Galileo Galilei
- 1988 Planetarium Copernicus
- 1992 Tellurium Johannes Kepler

1925
Pilot's Chronograph A.R.A. Navigacion

Case: silver, silver cuvette, leather strap, Ø 39 mm
Movement: nickel-plated, côtes de Genève, column-wheel chronograph control, manual
Remarks: very early crown-button chronograph with 30-minute counter; enamel dial with obvious origins in pocket watch style; numerals inlaid with luminous substance; two soldered silver wires serve as strap lugs; hinged case back
Estimated value: $7,800 ➔

1920
Chronograph

Case: 18-karat yellow gold, gold cuvette, leather strap, Ø 38 mm
Movement: nickel-plated, côtes de Genève, column-wheel control of chronograph, manual winding
Remarks: very early crown-button chronograph with 30-minute counter; enamel dial; hinged case back; two pocket watch bows serve as strap lugs
Estimated value: $3,900 ➔

1930
Chronograph

Case: yellow gold, gold cuvette, leather strap, Ø 34 mm
Movement: nickel-plated, côtes de Genève, column-wheel control of chronograph, manual winding
Remarks: very early crown-button chronograph with 30-minute counter and double tachymeter scale; outer tachymeter scale covers speeds to 60 km/h; scale in center of enamel dial covers speed measured to 10 km/h
Estimated value: $3,900 ➔

1925
Chronograph

Case: silver, silver cuvette, leather strap, Ø 38 mm
Movement: nickel-plated, column-wheel control of chronograph, manual winding
Remarks: very early wristwatch chronograph with 30-minute counter; enamel dial; the button for the chronograph functions is integrated into the crown; across from the crown is a second bow used for lugs; hinged case back
Estimated value: $3,900 ➔

1925
Chronograph

Case: 18-karat yellow gold, push-down case back, leather strap, Ø 34 mm
Movement: nickel-plated, column-wheel control of chronograph, manual winding
Remarks: elegant one button chronograph with 30-minute counter and additional pulsometer scale for easier pulse-reading; enamel dial; comparatively small subdials
Estimated value: $6,500 ➔

1925
Chronograph

Case: 18-karat yellow gold, push-down case back, leather strap, Ø 37 mm
Movement: gold-plated, frosted finish, column-wheel control of chronograph, manual winding
Remarks: early crown button chronograph with 30-minute counter and tachymeter scale; tachymeter scale snail on perimeter of enamel dial allows speed measurements to 20 km/h
Estimated value: $6,500 ➔

1930
Chronograph

Case: stainless steel, push-down case back, leather strap, 35 x 41 mm
Movement: nickel-plated, column-wheel control of chronograph, manual winding
Remarks: fine, early one-button chronograph with 30 minute counter at 3 o'clock; cushion-shaped stainless steel case; enamel dial
Estimated value: $4,550 ➔

1925
Chrongrafo Medical

Case: stainless steel, push-down case back, leather strap, Ø 37 mm
Movement: nickel-plated, column-wheel control of chronograph, manual winding
Remarks: early one-button chronograph with 30-minute counter and pulsometer scale; enamel dial; movable lugs
Estimated value: $3,250 ➔

1930
Doctor's Chronograph

Case: 18-karat yellow gold, push-down case back, leather strap, 32 x 41 mm
Movement: rhodium-plated, côtes de Genève, column-wheel control of chronograph, manual winding
Remarks: fine doctor's chronograph with 30-minute counter and pulsometer scale
Estimated value: $7,800 ➔

1938
Chronograph

Case: 18-karat yellow gold, push-down case back, leather strap, Ø 37 mm
Movement: nickel-plated, column-wheel control of chronograph, manual winding
Remarks: chronograph with 30-minute counter at 3 o'clock; for South American market; moveable lugs are especially striking
Estimated value: $4,550 ➔

1940
Chronograph

Case: stainless steel, push-down case back, leather strap, Ø 38 mm
Movement: rhodium-plated, côtes de Genève, column-wheel control of chronograph, manual winding
Remarks: chronograph with 30-minute and 12-hour counters
Estimated value: $3,900 ➔

1948
Chronograph

Case: stainless steel, screw-down case back, leather strap, Ø 35 mm
Movement: nickel-plated, column-wheel control of chronograph, manual winding
Remarks: chronograph with 30-minute counter at 3 o'clock; numerals inlaid with luminous substance for better legibility
Estimated value: $5,200 ➔

1945
Doctor's Chronograph

Case: 18-karat yellow gold, push-down case back, leather strap, Ø 37 mm
Movement: rhodium-plated, column-wheel control of chronograph, manual winding
Remarks: elegant one-button chronograph with 30-minute counter; pulsometer and tachymeter scales; covered lugs
Estimated value: $4,550 ➔

1935
Chronograph

Case: 18-karat yellow gold, push-down case back, leather strap, Ø 35 mm
Movement: rhodium-plated, côtes de Genève, column-wheel control of chronograph, manual winding
Remarks: very elegant one-button chronograph with 30-minute counter; tachymeter and telemeter scales; gold case with covered lugs
Estimated value: $6,500 ➔

1945
Chronograph

Case: stainless steel, 35 mm, bipartite, screw-down case back, leather strap
Movement: rhodium-plated, column-wheel control of chronograph, fine matte steel chronograph components
Remarks: sporty chronograph with 30-minute counter and three-color dial; tachymeter and telemeter scales
Estimated value: $5,200 ➔

1945
Chronograph

Case: stainless steel, push-down case back, leather strap, Ø 37 mm
Movement: Caliber R23, rhodium-plated, fausses côtes decoration, finely polished and beveled steel chronograph parts, column-wheel control, glucydur screw balance with shock protection, manual winding
Remarks: rare and fine chronograph with a 30-minute counter
Estimated value: $2,080 ➔

1960
Chronograph with Complete Calendar

Case: stainless steel, screw-down case back, leather strap, Ø 36 mm
Movement: Caliber N13HQ, nickel-plated, côtes de Genève, column-wheel control of chronograph, manual winding
Remarks: chronograph with 30-minute and 12-hour counters; complete calendar; sweep date hand; window displays for weekday and month
Estimated value: $5,200 ➔

1950
Chronograph with Complete Calendar

Case: 18-karat red gold, Ø 35 mm, tripartite, push-down case back, leather strap
Movement: red gold-plated, column-wheel control of chronograph, finely finished steel chronograph components
Remarks: fine red gold chronograph with 30-minute and 12-hour counters; complete calendar with date hand; window displays for weekday and month
Estimated value: $5,200 ➔

1943
Pilot's Watch

Case: stainless steel, screw-down case back, leather strap, Ø 38 mm
Movement: Caliber Felsa 264, rhodium-plated, polished, indirect sweep second, glucydur screw balance, manual winding
Remarks: very rare pilot's watch with indirect sweep seconds
Estimated value: $7,800 ➔

1970
Chronograph

Case: stainless steel, screw-down case back, leather strap, Ø 38 mm
Movement: Caliber N13B, gold-plated, côtes de Genève, column-wheel control of chronograph, manual winding
Remarks: heavy chronograph with 30-minute and 12 hour counters; tachymeter scale on bezel
Estimated value: $3,900 ➔

1980
Automatic Chronometer

Case: stainless steel, screw-down case back, leather strap, Ø 36 mm
Movement: Caliber NB1100, nickel-plated, 25 jewels, automatic winding
Remarks: wristwatch chronometer with ultra-fast frequency of 36,000 vph; date window at 3 o'clock; this watch was offered with a small chronometer certificate
Estimated value: $1,820 ➔

1965
Complete Calendar

Case: 18-karat yellow gold, Ø 27 mm, push-down case back, leather strap
Movement: Caliber 31.3.071, gold-plated, fausses côtes decoration, mirror-polished screws
Remarks: small gold ladies' watch with complete calendar; hand displays of date, weekday, and month; moon phase display in window at 12 o'clock
Estimated value: $3,250 ➔

1980
Chronograph with Complete Calendar

Case: 18-karat yellow gold, Ø 39 mm, tripartite, exhibition case back, leather strap
Movement: Valjoux Caliber 88, gold-plated, fausses côtes decorations, column-wheel control of chronograph, highly polished screws, fine matte steel chronograph components
Remarks: fine gold chronograph with 30-minute and 12-hour counters; complete calendar with date hand; window displays of weekday and month; moon phase at 6
Estimated value: $5,200 ➔

1989
Astrolabium Galileo Galilei

Reference number: 961-22
Case: 18-karat yellow gold, engraved bezel, screw-down transparent case back, leather strap, Ø 40 mm
Movement: Caliber 91.7.144, engraved, glucydur balance, automatic winding
Remarks: elegant, heavy astronomical wristwatch with a 24-hour display, month, weekday, with perpetual Gregorian calendar, zodiac, sun, moon, and star positions, solar and lunar eclipse display
Estimated value: $19,500 ➔

UNIVERSAL GENÈVE

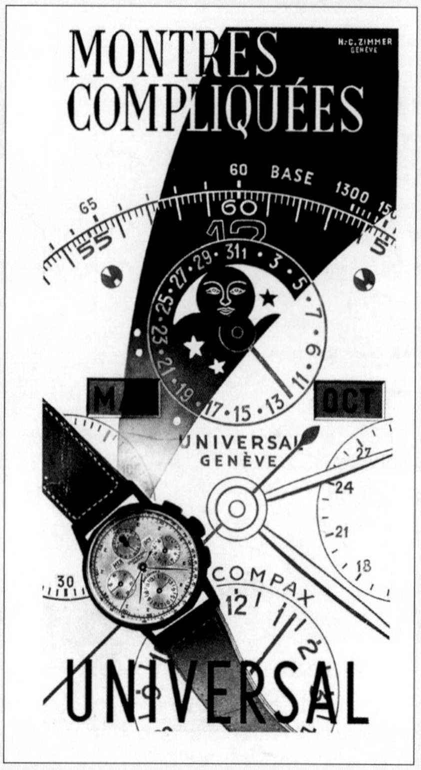

Like many brands that specialize in precise chronographs, Universal Genève's roots can be traced back to Le Locle. It was there that Numa Emile Descombes and Georges Perret founded the company Descombes & Perret in 1894. The goal they set was the development of high-quality watches with complications. But the pair did not want to be limited to just the design and manufacture of movements either; they also wanted to produce cases and dials themselves.

In the year of the company's founding, they applied for a patent for a watch with a combined 12- and 24-hour display, the additional 24-hour indication being actually a jump hour displayed in a window. That same year, the owners registered the brand Universal Watch.

The partnership came to an abrupt end when Descombes died in 1897. Perret had to look around for a new partner, whom he finally found in the young watchmaker Louis Berthoud.

In 1917, Universal Watch produced a large wrist chronograph with a 17-line movement already outfitted with an additional hour counter. This chronograph served at the front during World War I.

After the war, to make opening new markets a little easier, the owners moved their headquarters from Le Locle to Geneva, for in the big city there were much better travel options. Consequently, they changed the name of their company to Universal Genève.

One of the greatest achievements of this company was a patented hammer automatic from 1925, which was quite similar to Harwood's. Back then, Universal Genève specialized in chronographs whose excellent quality paid off in the 1930s when the company—like many others—was hit by the economic crisis. Because of their high production standards, they found financial backers to secure the company's continued existence. Thus, in 1941 they founded an additional company in Les Ponts-de-Martel in the Swiss Jura called Martel Watch.

Universal Genève's early success was due to the brand's focus on chronographs with two buttons for adding time measurements. In the 1930s, the model families Compur and Compax joined the existing Standard line.

Universal Genève introduced its best-known model in 1944: the Tri-Compax Chronograph, whose movement was additionally outfitted with a complete calendar and a moon phase display.

Universal Genève wrote aviation history with the Polerouter. SAS pilots wore them on their wrists when they opened lines from Scandinavia to Alaska and California by crossing the North Pole starting in 1954. To make sure that the Polerouter also worked correctly above the North Pole with its strong magnetic field, the watchmakers protected the movement with a soft iron core inside the case.

After 1967, Universal Genève partnered with Bulova and manufactured a watch using the Accutron Caliber 218. The cooperative venture ended with a takeover by Bulova. In 1977, Universal Genève once again became independent through a management buyout, but was sold to the Stelux Group nine years later.

Today, Universal Genève still manufactures a flat micro rotor caliber with excellent rate performance.

Founded:
1894 by Numa Emile Descombes and Ulysse Georges Perret in Le Locle

Headquarters:
Geneva

Status:
revived several times

Specialty:
chronographs, pilot's watches, diving watches

Significant inventions:
- 1925 two patents for automatic winding
- 1927 Convertible with a reversible case
- 1944 Tri-Compax
- 1954 Polerouter
- 1957 Caliber 215 Microtor

Leading models:
Polerouter, Tri-Compax, Aero-Compax

1935
8 Days

Case: chrome-plated, push-down case back, leather strap, 26 x 43 mm
Movement: gold-plated, frosted finish, manual winding
Remarks: rare men's watch with eight-day shaped movement; twin spring barrels
Estimated value: $3,900 ➔

1945
Chonograph

Case: stainless steel, push-down case back, leather strap, 26 x 35 mm
Movement: gold-plated, frosted finish, column-wheel control of chronograph, manual winding
Remarks: simple chronograph in square case
Estimated value: $2,600 ➔

1930
Chronograph

Case: 18-karat yellow gold, push-down case back, leather strap, 35 x 35 mm
Movement: gold-plated, frosted finish, column-wheel control of chronograph, finely finished chronograph steel components, polished screws, manual winding
Remarks: gold chronograph with cushion-shaped case and 45-minute counter, blue tachymeter and red telemeter scales
Estimated value: $5,850 ➔

1945
Calendar Watch with Moon Phase

Case: 18-karat white gold, push-down case back, leather strap, Ø 34 mm
Movement: Caliber 291, rhodium-plated, manual winding
Remarks: men's watch with complete calendar and moon phase; date at 3 o'clock; window display of weekday at 12 o'clock; moon phase and month at 6 o'clock; subsidiary seconds
Estimated value: $3,250 ➔

1940
Uni-Compax

Case: 18-karat rose gold, push-down case back, leather strap, 26 x 26 mm
Movement: Caliber 289, gold-plated, frosted finish, column-wheel control of chronograph, finely finished chronograph steel components, manual winding
Remarks: gold Uni-Compax chronograph in square case with teardrop-shaped strap lugs; the buttons are arranged parallel to the crown are noteworthy; the numbers 6 and 12 are especially emphasized on the dial
Estimated value: $4,550 ➔

1955
Automatic with Enamel Dial

Reference number: 100105 3
Case: 18-karat rose gold, bipartite, screw-down case back, leather strap, Ø 36 mm
Movement: Caliber 138 C, rhodium-plated, finely finished, pendulum oscillating weight, polished screws
Remarks: extremely rare gold watch with enamel dial in cloisonné technique; this watch was made for Saud ibn Abd al-Aziz, who was king of Saudi Arabia from 1953 to 1964
Estimated value: $7,150 ➔

1945
Aero-Compax

Case: stainless steel, push-down case back, leather strap, Ø 55 mm
Movement: Caliber 287, gold-plated, frosted finish, column-wheel control of chronograph, manual winding
Remarks: exceptionally large pilot's chronograph with 30-minute and 12-hour counters; settable control time in subdial at 12 o'clock
Estimated value: $10,400 ➔

1940
Split-Seconds Pilot's Chronograph for A. Cairelli

Case: stainless steel, push-down case back, leather strap, Ø 44 mm
Movement: rhodium-plated, double column-wheel control of chronograph, manual winding
Remarks: one-button pilot's chronograph with 24-hour dial, split-seconds function, and 16-minute counter for astronavigation; "AMI, Chronometro per Navigaz. Astronom. HA-I, N. Categ. 19620, MM. 200033" on case back
Estimated value: $19,500 ➔

1950
Aero-Compax

Case: stainless steel, push-down case back, leather strap, Ø 37 mm
Movement: nickel-plated, column-wheel control of chronograph, manual winding
Remarks: pilot's chronograph with 30-minute and 12-hour counters; settable control time in subdial at 12 o'clock
Estimated value: $3,900 ➔

1950
Tri-Compax

Case: 18-karat yellow gold, push-down case back, leather strap, Ø 35 mm
Movement: Caliber 481, rhodium-plated, column-wheel control of chronograph, manual winding
Remarks: chronograph with 30-minute and 12-hour counters; complete calendar; date and moon phase are displayed on subdial at 12 o'clock; window display of weekday and month
Estimated value: $5,850 ➔

1948
Dato-Compax

Case: 18-karat yellow gold, push-down case back, leather strap, Ø 35 mm
Movement: Caliber 285, gold-plated, frosted finish, column-wheel control of chronograph, manual winding
Remarks: gold Dato-Compax chronograph with 30-minute and 12-hour counters; date at 12 o'clock
Estimated value: $3,900 ➔

1955
Tri-Compax

Case: 18-karat yellow gold, push-down case back, leather strap, Ø 37 mm
Movement: Caliber 287, gold-plated, frosted finish, column-wheel control of chronograph, manual winding
Remarks: gold chronograph with 30-minute and 12-hour counters; complete calendar; date and moon phase are displayed on subdial at 12 o'clock; window display of weekday and month
Estimated value: $5,850 ➔

1945
Compax

Case: stainless steel, push-down case back, leather strap, Ø 34 mm
Movement: Caliber 285, gold-plated, frosted finish, column-wheel control of chronograph, manual winding
Remarks: chronograph with 30-minute and 12 hour counters
Estimated value: $1,820 ➔

1950
Uni-Compax

Case: stainless steel, push-down case back, leather strap, Ø 35 mm
Movement: Caliber 285, gold-plated, column-wheel control of chronograph, manual winding
Remarks: sporty elegant chronograph with 45-minute counter
Estimated value: $1,560 ➔

1940
Filmcompax

Case: stainless steel, push-down case back, leather strap, Ø 37 mm
Movement: Caliber 287, nickel-plated, column-wheel control of chronograph, manual winding
Remarks: chronograph with 30-minute and 12-hour counters; additional special measurement scale for directors and camera people on inner rotating bezel (flange)
Estimated value: $5,200 ➔

1960
Tri-Compax

Case: 18-karat yellow gold, push-down case back, leather strap, Ø 37 mm
Movement: Caliber 281, rhodium-plated, finely finish, column-wheel control of chronograph, finely finished chronograph steel components, polished screws, 17 jewels, manual winding
Remarks: gold Tri-Compax with 30-minute and 12-hour counters; complete calendar; moon phase display; tachymeter scale on outer ring of dial
Estimated value: $5,200 ➔

1950
Compax Climate Proof

Case: 18-karat red gold, push-down case back, leather strap, Ø 34 mm
Movement: Caliber 481, nickel-plated, column-wheel control of chronograph, manual winding
Remarks: rare red gold compax chronograph with 30-minute and 12-hour counters
Estimated value: $3,250 ➔

1950
Compax

Case: 18-karat red gold, push-down case back, leather strap, Ø 32 mm
Movement: Caliber 281, gold-plated, frosted finish, column-wheel control of chronograph, manual winding
Remarks: fine red gold chronograph with 30-minute and 12-hour counters; this watch was offered in its original box; remarkable teardrop-shaped strap lugs
Estimated value: $3,900 ➔

1968
Aero-Compax

Case: stainless steel, screw-down case back, leather strap, Ø 40 mm
Movement: Caliber 130, nickel-plated, column-wheel control of chronograph, manual winding
Remarks: rare large chronograph in solid stainless steel case with 30-minute and 12-hour counters; rotating bezel
Estimated value: $2,340 ➔

1968
Tri-Compax

Case: stainless steel, screw-down case back, leather strap, Ø 37 mm
Movement: Caliber 281, nickel-plated, column-wheel control of chronograph, manual winding
Remarks: sporty tri-compax chronograph with 30-minute and 12-hour counters; complete calendar; date and moon phase are displayed on subdial at 12 o'clock; window displays of weekday and month; tachymeter scale on the bezel
Estimated value: $4,550 ➔

1960
Tri-Compax

Case: stainless steel, screw-down case back, leather strap, Ø 31 mm
Movement: Caliber 281, rhodium-plated, column-wheel control of chronograph, manual winding
Remarks: sporty tri-compax chronograph with 30-minute and 12-hour counters; complete calendar; date and moon phase are displayed on subdial at 12 o'clock; window display of weekday and month
Estimated value: $3,900 ➔

1965
Polerouter Automatic

Case: stainless steel, screw-down case back, leather strap, Ø 34 mm
Movement: Caliber 215, nickel-plated, côtes de Genève, micro rotor, 28 jewels, automatic winding
Remarks: sporty men's watch with fine micro rotor automatic movement
Estimated value: $1,170 ➔

1940
Compur

Case: 18-karat yellow gold, tripartite, push-down case back, leather strap, Ø 34 mm
Movement: Caliber 285, gold-plated, frosted finish, column-wheel control of chronograph, very finely finished chronograph steel components, polished screws, manual winding
Remarks: rare Compur chronograph with 45-minute counter and tachymeter scale; the watch has a double signature "Altesse-Universal Genève"; the Roman numerals VI and XII stand out on the dial alongside the simple line markers
Estimated value: $3,900 ➔

1970
Chronograph

Case: gold-plated, push-down case back, leather strap, Ø 38 mm
Movement: Venus Caliber 178, gold-plated, column-wheel control of chronograph, manual windings
Remarks: simple chronograph with 30-minute and 12-hour counters
Estimated value: $910 ➔

VACHERON CONSTANTIN

Vacheron Constantin can rest assured: it is the oldest *manufacture* in the history of the Swiss watch industry. Unlike many companies with venerable names that have been founded in recent years, this firm has uninterruptedly manufactured timepieces since its opening more than 250 years ago.

The brand's tenure began in 1755 when Jean-Marc Constantin established his company in Geneva. The customers for his demanding watches were mainly located in France. That such an important market had developed there was thanks to that country's spoiled and splendor-craving noble class—Louis XVI himself was an enthusiastic watch collector.

In 1789, this market literally fell apart overnight. The French Revolution had broken out, and over the next few years, the nobles were more occupied with simply saving themselves. Watches were no longer a topic, at least for the time being.

Constantin had trouble keeping his head above water. When he died in 1805 and his heirs took over the company, they saw no other chance but to search out a new, solvent partner. In 1819, they found just the right person in François Vacheron, the son of a rich textile and wheat dealer. He took over the business side of the company that would now go by Vacheron & Constantin.

The company's greatest success can be attributed to watchmaker George-Auguste Leschot, who designed a machine in 1839 that allowed a precision movement to be manufactured in a more economical manner. This first step toward machine production was a small revolution for the watchmaking industry.

The company added wristwatches to its collection in about 1910 with movements that were mainly supplied by LeCoultre. Right from the beginning, this company had a soft spot for jeweled watches, and along with elegant timepieces destined for daily use, the watchmakers also created models with a more avant-garde design: in the 1920s Vacheron & Constantin already made a design watch whose dial could be covered by a small Venetian blind.

Since the 1930s, there has also been a great selection of chronographs in the company's collection. Additionally, Vacheron & Constantin offered a wristwatch with a minute repeater, though it was only manufactured in small series.

Although the company crafted high-quality watches, the owners were not able to keep it financially independent. The outbreak of World War II brought another bout of financial trouble, and Charles Constantin had no other choice but to sell the majority of stock in 1940. The new owner was Georges Ketterer. Under Ketterer's management, Vacheron & Constantin was able to secure a position as a manufacturer of luxury watches. The lion's share of the movements utilized were not of the company's own production, but were purchased from Jaeger-LeCoultre in Le Sentier.

In 1955, in celebration of the company's 200th anniversary, Vacheron & Constantin introduced an ultra-flat wristwatch: its movement was a mere 1.64 millimeters high. This caliber continues to be one of the world's thinnest watch movements.

When the brand introduced its first sports watch in 1974, this seemed almost contrary to the image of the otherwise discreet gold watches it was famous for. The design of Model 222 was reminiscent of a porthole, and the stylized Maltese cross that the company had registered in 1880 as its logo could be found in numerous details on the case.

Founded:
1755 by Jean-Marc Constantin in Geneva

Headquarters:
Geneva

Status:
active without interruption (today Richemont Luxury Group)

Specialty:
chronometers, jewelry watches for ladies and men, extra-flat watches, repeaters, sports time measurements, toy watches

Significant inventions:
- 1811 Virgule escapement
- 1922 Microphone to control the rate (predecessor of a watch timing machine)

Leading models:
Chronomètre Royal, Overseas

1936
Men's Watch

Case: platinum, bipartite, hinged push-down case back, leather strap, 22 x 44 mm
Movement: shaped movement, rhodium-plated, finely finished, polished screws, 17 jewels, regulated in 4 positions
Remarks: extremely rare men's watch in an elongated tonneau-shaped platinum case; Art Deco dial with minute scale on perimeter of dial; this watch was offered with a Patek Philippe white gold buckle
Estimated value: $15,600 →

1915
Men's Watch

Case: 18-karat yellow gold, tripartite, hinged push-down case back, gold cuvette, leather strap, Ø 36 mm
Movement: gold-plated, frosted finish, blued screws
Remarks: rare early wristwatch whose case is clearly reminiscent of a pocket watch
Estimated value: $3,250 →

1930
Montre à Volets

Case: 18-karat white gold and yellow gold, leather strap, 26 x 33 mm
Movement: rhodium-plated, regulated in 5 positions, manual winding
Remarks: extremely rare men's watch with Venetian blind technology over the dial activated by a second crown at 9 o'clock; this watch was offered in its original box and with certificate
Estimated value: $32,500 →

1922
Men's Watch

Case: 18-karat yellow gold, push-down case back, leather strap, 31 x 37 mm
Movement: gold-plated, frosted finish, manual winding
Remarks: early gold men's wristwatch in a cushion-shaped case; the subsidiary seconds dial is recessed into the main dial
Estimated value: $3,250 →

1940
Men's Watch

Reference number: 24
Case: stainless steel, push-down case back, leather strap, Ø 30 mm
Movement: rhodium-plated, fausses côtes decoration, compensation screw balance, 17 jewels, adjusted in 3 positions, manual winding
Remarks: made for Andreas Huber of Urania, Munich
Estimated value: $5,200 →

1940
Men's Watch

Case: platinum, push-down case back, leather strap, 25 x 39 mm
Movement: tonneau-shaped, rhodium-plated, fausses côtes decoration, glucydur screw balance, manual winding
Remarks: unusually elegant platinum men's watch with diamond indexes
Estimated value: $5,200 →

1935
Men's Watch

Case: stainless steel, push-down case back, leather strap, 24 x 38 mm
Movement: rhodium-plated, fausses côtes decoration, compensation screw balance, manual winding
Remarks: unusually simple men's watch in stainless steel
Estimated value: $3,900 →

1938
Men's Watch

Reference number: 20
Case: stainless steel, push-down case back, leather strap, Ø 30 mm
Movement: rhodium-plated, côtes de Genève, manual winding
Remarks: rare men's wristwatch with large crown partially recessed into case
Estimated value: $4,550 →

1946
Chronograph

Case: stainless steel, push-down case back, leather strap, Ø 33 mm
Movement: rhodium-plated, fausses côtes decoration, finely polished and beveled steel chronograph parts, column-wheel control, glucydur screw balance, manual winding
Remarks: sporty men's watch with a chronograph, very rare in a stainless steel case
Estimated value: $26,000 ↗

1940
Chronograph

Case: stainless steel, tripartite, push-down case back, leather strap, Ø 34 mm
Movement: Caliber 434, rhodium-plated, fausses côtes decoration, column wheel control of chronograph, finely finished chronograph steel components, polished screws, manual winding
Remarks: rare chronograph with 30-minute counter and additional telemeter scale; this watch was offered with a Vacheron Constantin buckle
Estimated value: $39,000 →

1938
Chronograph

Case: 18-karat red gold, push-down case back, leather strap, Ø 36 mm
Movement: Caliber V434, rhodium-plated, côtes de Genève, column-wheel control of chronograph, manual winding
Remarks: fine chronograph with 30-minute counter at 3 o'clock; this watch was offered with VC certificate and in its original box
Estimated value: $36,400 →

1945
Chronograph

Case: stainless steel, push-down case back, gold crown, leather strap, Ø 36 mm
Movement: rhodium-plated, fausses côtes decoration, finely polished and beveled steel chronograph parts, column-wheel control, glucydur screw balance, manual winding
Remarks: sporty men's watch with chronograph
Estimated value: $19,500 →

1942
Chronograph

Case: 18-karat red gold, push-down case back, leather strap, Ø 34 mm
Movement: Caliber V434, nickel-plated, côtes de Genève, column-wheel control of chronograph, manual winding
Remarks: fine chronograph with 30-minute counter at 3 o'clock
Estimated value: $35,100 →

1947
Chronograph

Case: 18-karat yellow gold, push-down case back, leather strap, Ø 34 mm
Movement: Caliber V434, rhodium-plated, côtes de Genève, column-wheel control of chronograph, manual winding
Remarks: fine chronograph with 30-minute counter at 3 o'clock; this watch was offered with VC certificate and in its original box
Estimated value: $32,500 →

1939
Men's Watch

Reference number: 4085
Case: 18-karat yellow gold, push-down case back, leather strap, Ø 34 mm
Movement: Caliber V203, rhodium-plated, côtes de Genève, manual winding
Remarks: extremely rare rectangular men's wristwatch with very original strap lugs; this watch was offered in its oariginal box and an excerpt from Vacheron Constantin's master registry
Estimated value: $7,800 →

1945
Men's Wristwatch

Case: 18-karat red gold, push-down case back, leather strap, 23 x 39 mm
Movement: Caliber V435/3C, rhodium-plated, côtes de Genève, manual winding
Remarks: rare men's watch in a fitted rectangular case with a Vacheron & Constantin gold buckle; this watch was offered in its original box
Estimated value: $5,850 →

1937
Chronograph

Case: 18-karat yellow gold, tripartite, push-down case back, leather strap, Ø 35 mm
Movement: gold-plated, frosted finish, column wheel control of chronograph, very finely finished chronograph steel components, polished screws, manual winding
Remarks: extremely rare chronograph with 45-minute counter and additional tachymeter scale
Estimated value: $36,400 →

1943
Men's Watch

Case: 18-karat red gold, push-down case back, leather strap, Ø 33 mm
Movement: Caliber V454, nickel-plated, côtes de Genève, manual winding
Remarks: rare red gold men's watch with lavish cloisonné enamel dial; gold buckle
Estimated value: $39,000 →

1960
Men's Watch

Reference number: 4108
Case: 18-karat rose gold, bipartite, push-down case back, gold link bracelet, 25 x 33 mm
Movement: Caliber 1001, rhodium-plated, fausses côtes decoration, polished screws, 18 jewels, regulated in 8 positions
Remarks: rare men's watch in a delicate square case; this watch was offered with an original Vacheron & Constantin yellow gold bracelet
Estimated value: $5,850 →

1955
Men's Watch

Case: 18-karat yellow gold, push-down case back, leather strap, 31 x 31 mm
Movement: Caliber 477/1, rhodium-plated, fausses côtes decoration, pendulum oscillating weight, polished screws, 17 jewels, regulated in three positions
Remarks: rare platinum watch with case back engraving "Archbishop Samuel J. David, Toledo Ohio M.D. 8-26-52"
Estimated value: $5,200 →

1945
Men's Watch

Case: 18-karat yellow gold, push-down case back, leather strap, Ø 33 mm
Movement: Caliber 458/2B, rhodium-plated, côtes de Genève, manual winding
Remarks: very fine gold men's watch; this watch was offered in its original box
Estimated value: $4,550 →

1947
Men's Watch

Case: 18-karat red gold, push-down case back, leather strap, Ø 31 mm
Movement: Caliber V458, nickel-plated, côtes de Genève, 17 jewels, manual winding
Remarks: simple men's watch with remarkable strap lugs
Estimated value: $3,380 →

1944
Men's Watch

Case: 18-karat yellow gold, bipartite, push-down case back, leather strap
Movement: Caliber 466/3B, rhodium-plated, fausses côtes decoration, polished screws, 17 jewels, regulated in one position
Remarks: elegant men's watch with remarkable strap lugs
Estimated value: $2,600 →

1947
Men's Watch

Case: 18-karat yellow gold, push-down case back, leather strap, Ø 31 mm
Movement: Caliber 466/3B, rhodium-plated, côtes de Genève, manual winding
Remarks: discreet men's watch with indirect sweep seconds
Estimated value: $3,250 →

1940
Perpetual Calendar
Case: stainless steel, push-down case back, leather strap, Ø 35 mm
Movement: rhodium-plated, fausses côtes decoration, glucydur screw balance, manual winding
Remarks: extremely rare Geneva men's wristwatch with a perpetual calendar in a stainless steel case
Estimated value: $15,600 →

1955
Men's Watch
Case: 18-karat red gold, bipartite, push-down case back, 18-karat red gold link bracelet, Ø 38 mm
Movement: Caliber P453/3B, rhodium-plated, fausses côtes decoration, polished screws, 17 jewels, regulated in one position
Remarks: rare large men's watch with solid gold link bracelet
Estimated value: $5,200 →

1948
Men's Watch
Case: 18-karat red gold, push-down case back, leather strap, Ø 33 mm
Movement: Caliber V453, rhodium-plated, côtes de Genève, manual winding
Remarks: elegant men's watch
Estimated value: $3,250 →

1943
Men's Watch
Case: 18-karat yellow gold, push-down case back, leather strap, Ø 34 mm
Movement: Caliber P453/3C, rhodium-plated, côtes de Genève, manual winding
Remarks: discreet men's wristwatch with subsidiary seconds
Estimated value: $3,250 →

1940
Men's Watch
Case: stainless steel, tripartite, push-down case back, leather strap, Ø 34 mm
Movement: Caliber 453, rhodium-plated, fausses côtes decoration, polished screws
Remarks: elegant men's watch with subsidiary seconds and two-tone dial
Estimated value: $3,250 →

1952
Men's Watch
Case: 18-karat yellow gold, push-down case back, leather strap, Ø 34 mm
Movement: Caliber P453/3B, rhodium-plated, côtes de Genève, manual winding
Remarks: simple men's watch with subsidiary seconds
Estimated value: $3,900 →

1955
Men's Watch
Reference number: 4711
Case: 18-karat yellow gold, tripartite, push-down case back, leather strap, Ø 36 mm
Movement: Caliber P453/3B, rhodium-plated, fausses côtes decoration, polished screws, 17 jewels, regulated in one position
Remarks: elegant gold men's watch; this watch was offered with an 18-karat gold buckle
Estimated value: $3,900 →

1951
Men's Watch
Case: 18-karat yellow gold, push-down case back, leather strap, Ø 37 mm
Movement: Caliber P454/5B, rhodium-plated, côtes de Genève, manual winding
Remarks: elegant men's watch in heavy gold case
Estimated value: $5,200 →

1952
Men's Watch

Reference number: 4413
Case: 18-karat red gold, push-down case back, leather strap, Ø 35 mm
Movement: Caliber P454/5B, nickel-plated, côtes de Genève, manual winding
Remarks: simple men's watch; this watch was offered with original gold buckle
Estimated value: $2,990 ➔

1978
Extra Flat

Reference number: 6352
Case: 18-karat white gold, push-down case back, white gold Milanaise bracelet, Ø 32 mm
Movement: Caliber 1003, rhodium-plated, côtes de Genève, 17 jewels, regulated in eight positions, manual winding
Remarks: rare extra-flat men's wristwatch in white gold; the movement was adjusted in eight positions
Estimated value: $3,900 ➔

1969
Men's Watch

Case: 18-karat yellow gold, push-down case back, leather strap, Ø 34 mm
Movement: Caliber K1001, rhodium-plated, côtes de Genève, manual winding
Remarks: elegant men's watch
Estimated value: $2,340 ➔

1956
Men's Watch

Case: 18-karat yellow gold, push-down case back, leather strap, Ø 33 mm
Movement: Caliber 1001, rhodium-plated, côtes de Genève, manual winding
Remarks: elegant men's wristwatch in a fine gold case
Estimated value: $2,080 ➔

1965
Men's Watch

Case: 18-karat yellow gold, push-down case back, leather strap, Ø 33 mm
Movement: Caliber K1014, rhodium-plated, côtes de Genève, manual winding
Remarks: elegant men's watch in a fine gold case
Estimated value: $2,340 ➔

1965
Men's Watch with Stop-Seconds

Case: 18-karat red gold, push-down case back, leather strap, Ø 33 mm
Movement: Caliber P1008/BS, nickel-plated, côtes de Genève, Seal of Geneva, 19 jewels, manual winding
Remarks: rare men's wristwatch with hacking seconds, which makes setting the precise time easier; the movement was adjusted in eight positions
Estimated value: $3,900 ➔

1966
Chronomètre Royal

Case: 18-karat yellow gold, push-down case back, leather strap, Ø 34 mm
Movement: Caliber P1008/BS, rhodium-plated, côtes de Genève, Seal of Geneva, manual winding
Remarks: rare chronometer with subsidiary seconds; the movement hacks for easier setting of the precise time
Estimated value: $8,450 ➔

1956
Men's Watch

Case: 18-karat yellow gold, push-down case back, leather strap, Ø 38 mm
Movement: Caliber P454/5B, rhodium-plated, côtes de Genève, 17 jewels, manual winding
Remarks: simple large men's watch
Estimated value: $3,900 ➔

1968
Chronomètre Royal

Reference number: 6694
Case: 18-karat yellow gold, screw-down case back, leather strap, 35 x 40 mm
Movement: Caliber 1072/1, rhodium-plated, côtes de Genève, Seal of Geneva, 29 jewels, regulated in eight positions, automatic winding
Remarks: chronometer with date window at 3; offered with additional original yellow link bracelet; movement with gold rotor on jewel bearings adjusted in 8 positions
Estimated value: $9,100 →

1960
Automatic Men's Watch

Reference number: 6038
Case: 18-karat yellow gold, screw-down case back, leather strap, Ø 35 mm
Movement: Caliber P1019, rhodium-plated, côtes de Genève, 21 jewels, regulated in 8 positions, automatic winding
Remarks: discreet automatic men's wristwatch with 18-karat gold rotor on jewel bearings; the movement was regulated in eight positions
Estimated value: $4,550 →

1960
Men's Watch

Case: 18-karat yellow gold, screw-down case back, leather strap, Ø 35 mm
Movement: Caliber P1019, rhodium-plated, côtes de Genève, automatic winding
Remarks: simple gold men's watch
Estimated value: $3,900 →

1969
Men's Watch

Reference number: 6562
Case: stainless steel, bipartite, screw-down case back, leather strap, Ø 35 mm
Movement: Caliber K1072, rhodium-plated, fausses côtes decoration, 18-karat gold rotor in jeweled bearings, polished screws, Seal of Geneva, 29 jewels, regulated in 8 positions
Remarks: elegant men's watch with date at 3 o'clock
Estimated value: $3,900 →

1943
Automatic Men's Watch

Case: 18-karat yellow gold, screw-down case back, leather strap, Ø 35 mm
Movement: Caliber V477, nickel-plated, côtes de Genève, automatic winding
Remarks: extremely rare men's watch in 18-karat gold; the movement has a jewel-bearing hammer or pendulum for the automatic winding
Estimated value: $4,550 →

1951
Automatic Men's Watch

Case: 18-karat yellow gold, screw-down case back, leather strap, Ø 36 mm
Movement: Caliber P1019, rhodium-plated, côtes de Genève, automatic winding
Remarks: elegant automatic men's watch
Estimated value: $4,290 →

1965
Automatic Men's Watch

Reference number: 6378
Case: 18-karat yellow gold, screw-down case back, leather strap, Ø 35 mm
Movement: Caliber K1071, rhodium-plated, côtes de Genève, 29 jewels, regulated in 8 positions, automatic winding
Remarks: discreet automatic men's watch with gold rotor on jewel bearings; the movement was regulated in eight positions
Estimated value: $4,550 →

1967
Automatic Men's Watch

Reference number: 6782
Case: 18-karat yellow gold, screw-down case back, rotating bezel, leather strap, Ø 36 mm
Movement: Caliber K1072, rhodium-plated, côtes de Genève, Seal of Geneva, 29 jewels, regulated in 8 positions, automatic winding
Remarks: extremely rare men's watch with rotating bezel and date display in window at 3 o'clock; the movement was regulated in eight positions
Estimated value: $5,850 →

1969
Automatic Men's Watch
Reference number: 6394
Case: 18-karat white gold, screw-down case back, leather strap, Ø 35 mm
Movement: Caliber 1072/1, nickel-plated, côtes de Genève, automatic winding
Remarks: automatic men's watch in a classic white gold case with date display at 3 o'clock; gold rotor on jewel bearings; this watch was originally sold in Okinawa
Estimated value: $7,150 ➔

1969
Automatic Men's Watch
Reference number: 7397
Case: 18-karat white gold, screw-down case back, leather strap, Ø 35 mm
Movement: Caliber 1072/1, nickel-plated, côtes de Genève, automatic winding
Remarks: men's watch in white gold case with date display at 3 o'clock; gold rotor on jewel bearings; this watch was offered with its original gold buckle, sales tag, certificate, and instruction booklet; originally sold in Okinawa
Estimated value: $7,150 ➔

1968
Automatic Men's Watch
Reference number: 6727
Case: 18-karat yellow gold, screw-down case back, leather strap, Ø 36 mm
Movement: Caliber 1071, nickel-plated, côtes de Genève, automatic winding
Remarks: very rare men's watch of which an estimated 500 pieces were made; gold rotor on jewel bearings; this watch was offered with its original gold buckle
Estimated value: $5,200 ➔

1969
Automatic Men's Watch
Reference number: 6732
Case: 18-karat yellow gold, screw-down case back, leather strap, Ø 35 mm
Movement: Caliber 1072/1, nickel-plated, côtes de Genève, automatic winding
Remarks: men's watch in gold case with date window at 3 o'clock; gold rotor on jewel bearings; this watch was offered with its original gold buckle
Estimated value: $4,550 ➔

1955
Chronomètre Royal
Case: 18-karat yellow gold, push-down case back, leather strap, Ø 35 mm
Movement: Caliber P1008/BS, rhodium-plated, fausses côtes decoration, glucydur screw balance, 19 jewels, manual winding
Remarks: chronometer with split-seconds system
Estimated value: $9,750 ➔

1961
Automatic
Reference number: 6440Q
Case: 18-karat yellow gold, screw-down case back, leather strap, Ø 36 mm
Movement: Caliber 1071/1, rhodium plated, fausses côtes decoration, glucydur screw balance, 29 jewels, adjusted in 8 positions, automatic winding
Estimated value: $5,200 ➔

1956
Men's Watch "Cioccolatone"
Reference number: 6440Q
Case: 18-karat yellow gold, push-down case back, leather strap, 36 x 44 mm
Movement: Caliber 1072, rhodium-plated, fausses côtes decoration, glucydur screw balance, 29 jewels, adjusted in 8 positions, automatic winding
Remarks: Italian collectors have nicknamed this square watch the "chocolate square"; new edition from 2003 under the name Toledo
Estimated value: $26,000 ➚

1990
Chronograph "Les Historiques"
Reference number: 47101
Case: 18-karat yellow gold, push-down case with transparent case back, leather strap, Ø 36 mm
Movement: Caliber 1140, rhodium-plated, fausses côtes decoration, finely polished and beveled steel chronograph parts, column-wheel control, screw balance, 21 jewels, adjusted in 5 positions, manual winding
Remarks: very fine men's watch with chronograph
Estimated value: $15,600 ➚

VULCAIN

Vulcan, the Roman god of fire, is described in mythology as an artistic smith. This is an entirely suitable name giver for a watch company.

The company had to be renamed several times to finally end up with this moniker, though: In 1858 it was registered in La Chaux-de-Fonds as Maurice Ditisheim. Almost forty years later, the owners added the name "Fabrique Vulcain" to what was essentially the last name of the company's founder. In the coming years, the name was repeatedly changed. From 1900 on, it was finally called Vulcain—to be renamed a decade later as Ditisheim & Cie. In 1911, it received the predicate "Fabrique Vulcain & Volta."

Hardly another watch brand is as strongly associated with one single product as Vulcain is with its Cricket. In 1947, Vulcain introduced its first alarm wristwatch. Despite its smaller case dimensions, it was as loud as a cricket thanks to the doubly thick membrane case back, which is what gave it its name.

The Cricket sold so well through the decades that over the course of time, it became a synonym for the Vulcain brand. Its most famous wearers were in the United States where it remained "time minister" for many decades on the wrist of the president.

Alongside the Cricket, Vulcain also had other watches in its line—most of which were outfitted with the brand's own calibers.

In 1961, the company was restructured once more: Vulcain SA became a founding member of MSR Holding (Manufactures Suisses Réunies), whose main stockholder was the company Revue. At the end of the 1970s, it was decided that from then on all watches of the MSR Group would only be distributed under the brand name Revue Thommen—except the Cricket. It remained known under the old name Vulcain Cricket until 1986, but had to bow to the Revue Thommen brand as well at that time.

Revue sold the rights to the brand name Vulcain SA around the turn of the millennium. Soon, though, the brand was in trouble again and had to undergo yet another restructuring. In the fall of 2001, PMH purchased all rights to the name with the goal of resuscitating the old Vulcain brand and relaunching the Cricket. PMH, in turn, sold it to the Swiss retail chain Les Ambassadeurs in late 2009 after launching a Cricket version based on a new autonomous caliber. And so the fire continues to burn.

Founded:
1858 by Maurice and Gaspard Ditisheim in La Chaux-de-Fonds

Headquarters:
Le Locle

Status:
active (Excellence Holding AG)

Specialty:
chronometers, repeaters, wristwatch alarms, diving watches

Significant inventions:
- 1942 Cricket
- 1942 Exactomatic
- 1961 Nautical with a decompression calculator

Leading models:
Cricket, Nautical

1925
Men's Watch

Case: silver, push-down case back, leather strap, 25 x 40 mm
Movement: rhodium-plated, côtes de Genève, manual winding
Remarks: fine men's watch in unusual oval case
Estimated value: $780 ➔

1940
Régulateur

Case: stainless steel, push-down case back, rotating bezel, leather strap, Ø 32 mm
Movement: rhodium-plated, manual winding
Remarks: men's watch with regulator dial
Estimated value: $390 ➔

1960
Cricket

Case: 18-karat red gold, double push-down case back, leather strap, Ø 38 mm
Movement: Vulcain Caliber 120, nickel-plated, twin spring barrels, 17 jewels, manual winding
Remarks: gold alarm wristwatch with double sounding case back; this watch was offered in its original box
Estimated value: $2,080 ➔

1965
Cricket Calendar

Case: stainless steel, double push-down case back, leather strap, Ø 34 mm
Movement: Vulcain Caliber 402 (S1), nickel-plated, 17 jewels, manual winding
Remarks: alarm wristwatch with subsidiary seconds and date; movement with spring barrel for movement and alarm
Estimated value: $1,040 ➔

1965
Chronograph with Complete Calendar

Case: chrome-plated, stainless steel screw-down case back, leather strap, Ø 39 mm
Movement: Vulcain Caliber 730, rhodium-plated, column-wheel control of chronograph, manual winding
Remarks: chronograph with 30-minute and 12-hour counters; complete calendar
Estimated value: $1,560 ➔

1956
Cricket Golden Voice

Case: gold-plated, stainless steel push-down case back, leather strap
Movement: Vulcain Caliber 406, rhodium-plated, Exactomatic, 17 jewels, manual winding
Remarks: small ladies' alarm in an early case version
Estimated value: $1,040 ➚

1961
Cricket Nautical

Case: stainless steel, double push-down case back, Tropic plastic strap, Ø 42 mm
Movement: Vulcain Caliber 120, nickel-plated, twin spring barrels, 17 jewels, manual winding
Remarks: extremely rare professional diver's alarm wristwatch; water-resistant to 300 meters; using length and depth of dive, the decompression time necessary can be seen in a cutaway in the dial
Estimated value: $10,400 ➔

1969
Cricket Nautical

Case: stainless steel, push-down case back, plastic strap, Ø 42 mm
Movement: Vulcain Caliber MSR S2, rhodium-plated, côtes de Genève, Exactomatic, 17 jewels, manual winding
Remarks: rare Cricket Nautical; the colorful rings on the dial display the various decompression times
Estimated value: $1,820 ➔

WAKMANN/ GIGANDET

What might seem like a double name for a venerable brand is in reality the double signature of two independent companies separated by the Atlantic Ocean.

Wakmann Watch Co., founded in New York, was the importer of various brands for the North American market. The most famous brand that it distributed was Breitling. Some of these watches even got a double signature and were known as Breitling Wakmann—although never Breitling's leading models.

Parallel to its distribution business after World War II, Wakmann began to produce—or have produced—watches under its own name.

Since the distribution of Breitling watches forced them to deal mainly with chronographs, it comes as no surprise that Wakmann focused mainly on creating chronographs as well. These were not produced in the United States, however, but purchased whole from suppliers in Switzerland. Charles Gigandet SA in Tramelan was the largest supplier for Wakmann Watch Co. Behind this company name, however, were hidden three different, independent manufacturers. Besides the company in Tramelan, there was also Gigandet-Rieder & Cie., which began its career in Langendorf but later moved to Solothurn, and the Gigandet & Chopard workshop, which was also located in Tramelan, high in the hills above Neuchâtel.

The watches that Gigandet made for Wakmann and itself under its own brand name were not surprisingly similar to Breitling models of the day. Some of the dials and cases were certainly consciously designed in the same vein, and possibly even purchased from the same suppliers.

Although Wakmann's watches were strictly "only" products of a trade name, their aesthetic and technical quality fulfilled the highest demands. Nothing was spared on the technical side above all, and thus Wakmann watches were outfitted with movements that are sought after today: leading the pack were various Valjoux chronograph movements with column-wheel control such as calibers 72, 88, and 730 as well as the rare Valjoux Caliber 92.

Alongside the previously mentioned models offered under the double signature "Breitling Wakmann," there were also individual models on the American market signed Wakmann Gigandet, even though these watches were normally exported to the United States only as Wakmann. There were identical models in Europe and in other markets with only Gigandet printed on the dial.

Wakmann also supplied the U.S. military with board clocks. These were not made in the company's own workshop either, but rather purchased from various manufacturers who then added the Wakmann name. Despite this, the mere fact that Wakmann supplied the army is proof of the quality of the company's private label watches.

Founded:
1947, founded together with Breitling as the Breitling Watch Corporation of America

Headquarters:
New York

Status:
dissolved

Specialty:
sale of imported watches, particularly for the Breitling brand

1970
Chronograph

Case: 18-karat yellow gold, push-down case back, leather strap, Ø 38 mm
Movement: Landeron Caliber 248, rhodium-plated, manual winding
Remarks: chronograph with 45-minute counter
Estimated value: $1,820 ➔

1965
Chronograph

Case: stainless steel, screw-down case back, leather strap, Ø 37 mm
Movement: nickel-plated, decorated, manual winding
Remarks: chronograph with 30-minute counter
Estimated value: $1,560 ➔

1975
Chronograph

Case: stainless steel, screw-down case back, leather strap, Ø 37 mm
Movement: Valjoux Caliber 726, rhodium-plated, column-wheel control of chronograph, manual winding
Remarks: chronograph with 30-minute and 12-hour counters
Estimated value: $1,690 ➔

1975
Chronograph with Complete Calendar

Case: stainless steel, screw-down case back, leather strap, Ø 37 mm
Movement: Valjoux Caliber 730, gold-plated, column-wheel control of chronograph, manual winding
Remarks: chronograph with 30-minute and 12-hour counters; complete calendar
Estimated value: $2,340 ➔

1970
Chronograph with Complete Calendar

Case: stainless steel, screw-down case back, leather strap, Ø 37 mm
Movement: Valjoux Caliber 730, gold-plated, column-wheel control of chronograph, manual winding
Remarks: chronograph with 30-minute and 12-hour counters; complete calendar
Estimated value: $2,340 ➔

1965
Chronograph with Complete Calendar

Case: stainless steel, screw-down case back, leather strap, Ø 37 mm
Movement: Valjoux Caliber 723, gold-plated, column-wheel control of chronograph, manual winding
Remarks: chronograph with 30-minute and 12-hour counters; complete calendar
Estimated value: $2,340 ➔

1972
Chronograph

Case: stainless steel, screw-down case back, rotating bezel, leather strap, Ø 38 mm
Movement: Landeron Caliber 51, nickel-plated, manual winding
Remarks: chronograph with 45-minute counter
Estimated value: $780 ➔

1975
Automatic Chronograph Complete Calendar

Case: stainless steel, push-down case back, leather strap, Ø 42 mm
Movement: Lemania Caliber LWO 1341, nickel-plated, automatic winding
Remarks: automatic chronograph with sweep 60-minute and off-center 12-hour counters; date, weekday; additional regatta scale
Estimated value: $1,040 ➔

ZENITH

FABRIQUES DES MONTRES ZENITH • LE LOCLE (Suisse)
VUE D'ENSEMBLE

With the introduction of the world's first automatic chronograph caliber Zenith secured a place of honor in the history of Swiss watchmaking. In 1865, twenty-two-year-old Georges Favre-Jacot founded his own workshop in Le Locle. He wanted to establish a more rational production for precise pocket watches. Favre-Jacot had recognized that it was impossible to meet the growing demand for export using the usual manufacturing methods. Until then, machine-produced components still needed to be individually fit by hand. The effort was too great and the quality he achieved—at least on standard movements—was too poor to compete on the international market.

In the main export country—the United States—manufacturers like Waltham and Elgin produced better watches than the Swiss even at their lowest quality level. This is where Favre-Jacot saw his big chance: if he were to manufacture according to American methods, he would logically also find success with his watches. And he was right; just ten years later more than a third of all employees in Le Locle were engaged in his enterprise, with his little workshop long since converted into a factory.

The watchmaker proved the quality of his watches convincingly when he won the first gold medal at the Swiss national exhibition in Geneva in 1896. Numerous other awards and many wins at chronometer competitions followed. Thanks to the precision of his watches, Favre-Jacot was also able to stand up to his competition when the Italian railway sought bids for a large contract for watches for its employees in 1909. The Zenith pocket watch became so successful, that the entire company was soon called Zenith.

In 1911, Favre-Jacot retired, leaving his son-in-law James Favre in charge of the company. He recognized the wristwatch trend in time and developed his own *manufacture* movements accordingly. After World War II, the first automatic watches arrived, still outfitted with hammer automatic winding.

Zenith had begun manufacturing chronographs in the 1920s. At first the company used pocket watch calibers, but later they put movements by chronograph specialist Excelsior Park into their cases. Zenith only took on its own movement production in 1960 when it acquired Martel Watch—Universal Genève's production workshop. Universal Caliber 285 became Zenith Caliber 146, which the company used to outfit pilot's chronographs in the early 1970s. The company manufactured these watches, called Cronometro Tipo CP2, exclusively for A. Cairelli in Rome.

In 1969, Zenith once again captured the industry's attention, introducing the first chronograph movement with automatic winding to the world. They just beat out the developmental group comprising Heuer, Breitling, Hamilton-Büren, and Dubois Dépraz and thus proudly named their chronograph El Primero. "The first" was not only outfitted with a classic column wheel but also an extremely fast oscillation rate of 36,000 vibrations per hour.

That this caliber is still in production today is due to the insight of one visionary employee: in the late 1970s he did not sell off the tooling for the erstwhile unprofitable mechanical caliber as directed, but stored it in the attic of the manufacture. The watch company Zenith was taken over by the American Zenith Radio Company, which later sold it to Swiss machine maker Dixi. Thanks to its movement know-how, today Zenith is the crown jewel in the watch division of luxury goods concern Louis Vuitton, Moët & Hennessy (LVMH).

Founded:
1865 by Georges Favre-Jacot in Le Locle

Headquarters:
Le Locle

Status:
active without interruption (today LVMH)

Specialty:
chronometers, chronographs, automatic watches

Significant discoveries:
1969 first automatic chronograph caliber El Primero

Leading models:
El Primero chronographs

1925

Chronograph

Case: sterling silver, push-down case back, leather strap, Ø 34 mm
Movement: rhodium-plated, côtes de Genève, column-wheel control of chronograph, manual winding
Remarks: early crown-button chronograph with 30-minute counter; enamel dial
Estimated value: $1,690 ➔

1940

Chronograph Compax

Case: 18-karat yellow gold, push-down case back, leather strap, Ø 34 mm
Movement: Caliber 146, rhodium-plated, côtes de Genève, column-wheel control of chronograph, manual winding
Remarks: chronograph with 30-minute and 12-hour counters
Estimated value: $2,080 ➔

1940

Chronograph

Case: stainless steel, screw-down case back, leather strap, Ø 37 mm
Movement: Caliber 143-6, nickel-plated, column-wheel control of chronograph, manual winding
Remarks: chronograph with 45-minute counter
Estimated value: $1,300 ➔

1965

Chronograph

Case: stainless steel, screw-down case back, leather strap, Ø 38 mm
Movement: Caliber 146HP, nickel-plated, column-wheel control of chronograph, manual winding
Remarks: sporty steel chronograph with 30-minute and 12-hour counters
Estimated value: $780 ➔

1965

Port Royal Chronometer Cal. 135

Case: stainless steel, push-down case back, leather strap, Ø 36 mm
Movement: Caliber 135, rhodium-plated, côtes de Genève, beveled, 19 jewels, manual winding
Remarks: chronometer wristwatch with unusual dial design
Estimated value: $4,550 ➔

1960

Chronometer Cal. 135

Case: stainless steel, screw-down case back, leather strap, Ø 36 mm
Movement: Caliber 135, rhodium-plated, with côtes de Genève, beveled, 19 jewels, manual winding
Remarks: rare chronometer wristwatch
Estimated value: $3,900 ➚

1965

Chronometer 2000 Cal. 135

Case: stainless steel, screw-down case back, leather strap, Ø 35 mm
Movement: Caliber 135, rhodium-plated, with côtes de Genève, beveled, 19 jewels, manual winding
Remarks: chronometer wristwatch
Estimated value: $3,640 ➔

1965

Chronometer

Case: stainless steel, screw-down case back, leather strap, Ø 334 mm
Movement: Caliber 40-T, rhodium-plated, beveled, manual winding
Remarks: rare chronometer wristwatch
Estimated value: $1,690 ➔

1965
Chronomètre Captain de Luxe

Case: 18-karat yellow gold, screw-down case back, leather strap, Ø 37 mm
Movement: Caliber 2562PC, nickel-plated, automatic winding
Remarks: rare gold automatic chronometer
Estimated value: $2,600 →

1966
Chronomètre Captain Automatic

Case: red gold, screw-down case back, leather strap, Ø 35 mm
Movement: Caliber 133.8, nickel-plated, côtes de Genève, automatic winding
Remarks: chronometer wristwatch with hammer automatic winding, manufactured in an estimated edition of 200 pieces; red gold buckle
Estimated value: $3,250 →

1970
Pilot's Chronograph Tipo CP-2

Case: stainless steel, screw-down case back, rotating bezel, leather strap, Ø 43 mm
Movement: Caliber 146DP, nickel-plated, column-wheel control of chronograph, manual winding
Remarks: pilot's chronograph with 30-minute counter; this watch bears the additional signature of the contractor: "A. Cairelli Roma Tipo CP-2"
Estimated value: $3,250 →

1975
Chronograph signed Türler

Case: stainless steel, screw-down case back, rotating bezel, leather strap, Ø 40 mm
Movement: Caliber 146HP, nickel-plated, column-wheel control of chronograph, manual winding
Remarks: chronograph with 30-minute and 12-hour counters; this watch was made for Türler in Switzerland and the dial is printed with its logo
Estimated value: $2,080 →

1975
Chronograph El Primero

Case: stainless steel, screw-down case back, stainless steel link bracelet, 38 x 41 mm
Movement: Caliber 3019PHC, nickel-plated, column-wheel control of chronograph, automatic winding
Remarks: chronograph with 30-minute and 12-hour counters; El Primero stands for the first serially manufactured automatic chronograph
Estimated value: $1,950 →

1975
Chronograph El Primero

Case: stainless steel, push-down case back, stainless steel link bracelet, 38 x 41 mm
Movement: Caliber 3019PHC, nickel-plated, column-wheel control of chronograph, 31 jewels, automatic winding
Remarks: chronograph with 30-minute and 12-hour counters; El Primero stands for the first serially manufactured automatic chronograph
Estimated value: $1,950 →

1972
Pilot's Chronograph El Primero

Case: stainless steel, screw-down case back, rotating bezel, leather strap, Ø 44 mm
Movement: Caliber 3019PHC, rhodium-plated, column-wheel control of chronograph, automatic winding
Remarks: rare pilot's chronograph of which only approximately 2,700 were made
Estimated value: $3,250 →

1975
Chronograph El Primero

Case: stainless steel, screw-down case back, stainless steel link bracelet, 42 x 44 mm
Movement: Caliber 3019PHC, nickel-plated, column-wheel control of chronograph, 31 jewels, automatic winding
Remarks: former leading model of the early El Primero chronograph line; a sketch of this model long graced Zenith's stationery, remarkable buttons designed as a rocking bar
Estimated value: $3,900 →

ZODIAC

One special element of the brand Zodiac was a certain loyalty to its name in the shape of the zodiac signs to which various models were assigned. The symbols in question were usually to be found on the back of each watch concerned.
This company was founded in 1882 by Ariste Calame. The Zodiac trademark—which came later—was registered in 1899. The company was soon after named Montres Zodiac. And in 1900, the founder withdrew and left the reins of the company to his son, which is when the assembly operation shifted to making movements.
In 1930, Zodiac presented its own shock protection system, and in 1937 the brand came out with a tonneau-shaped watch with an eight-day power reserve. This was followed in 1948 by the Astrographic, the brand's first automatic watch. At first, it was manufactured using a hammer automatic, but ensuing models were produced with rotor-driven automatic calibers manufactured by AS. The Astrographic's specialty was a power reserve display indicated by one of the hands, which showed up on the edge of the dial at 6 or 12 o'clock depending on the caliber used.
The Sea Wolf model represented the company's own diver's watch. U.S. Navy Demolition Teams put it to good use in the 1960s. The Super Sea Wolf, presented in 1968, was water-resistant to 750 meters at first, later to 1,000 meters.
The Pharaon line presented in 1958, and available both as a wristwatch and a pendant watch, contained hour markers in the shape of zodiac symbols. The specialties of this brand included ultra-flat wristwatches such as the wristwatch chronometers with subsidiary seconds introduced in 1959.
Ten years later, the new version of the Astrographic model came on the market. Its time display was achieved by two markings on plastic (acetate) disks, which left a mysterious impression. This model was also offered in a version outfitted with the SST caliber introduced in 1970, this same year Zodiac presented Caliber SST as a high-frequency movement beating at 36,000 vph.
In 1982, half of the company's stock was taken over by Dixi in Le Locle. After Dixi pulled out, the long-established company went bankrupt in 1997. Since 2002, this brand has belonged to the Fossil Group.

Founded:
1882 by Ariste Calame in Le Locle

Headquarters:
Le Locle

Status:
revived (called Fossil Inc. today)

Specialty:
flat men's watches, automatic watches, diving watches, chronometer

Significant inventions:
- 1958 Sea Wolf diving watches
- 1960 special pressing of the case back seal to increase water resistance
- 1968 Super Sea Wolf
- 1969 Astrographic

Leading models:
Sea Wolf, Astrographic

1955
Chronograph

Case: stainless steel, bipartite, screw-down case back, leather strap, Ø 35 mm
Movement: Valjoux Caliber 92, silver-plated, column-wheel control of chronograph, 17 jewels, manual winding
Remarks: sporty chronograph with 30-minute counter on the basis of the Valjoux Caliber 92; noticeable is the three-minute division for the minute hand at 3 o'clock
Estimated value: $650 ➔

1955
Extra Flat

Case: gold-plated, push-down case back, leather strap, Ø 33 mm
Movement: Zodiac Caliber 52, rhodium-plated, finely finished, additional shock protection for the escape wheel
Remarks: extra-flat wristwatch in a plated case with no second hand
Estimated value: $105 ➔

1955
Chronomètre Automatic

Case: 18-karat yellow gold, bipartite, push-down case back, leather strap
Movement: AS Caliber 1628, rhodium-plated, 17 jewels, regulated in seven positions, automatic winding
Remarks: gold men's watch with chronometer certificate
Estimated value: $1,040 ➔

1970
Automatic

Reference number: 762.935
Case: stainless steel, bipartite, push-down case back, leather strap, 35 x 34.5 mm
Movement: Caliber 76, nickel-plated, finely finished, special fine adjustment through spring loaded indirect regulator, 17 jewels, automatic winding
Estimated value: $200 ➔

GLOSSARY

ANNUAL CALENDAR

The automatic allowances for the different lengths of each month of a year in the calendar module of a watch. This type of watch usually shows the month and date, and sometimes the day of the week (like this one by Patek Philippe) and the phases of the moon.

ANTIMAGNETIC

Magnetic fields found in common everyday places affect mechanical movements, hence the use of anti- or non-magnetic components in the movement. Some companies encase movements in antimagnetic cores such as Sinn's Model 756, the Duograph, shown here.

ANTIREFLECTION

A film created by steaming the crystal to eliminate light reflection and improve legibility. Antireflection functions best when applied to both sides of the crystal, but because it scratches, some manufacturers prefer to have it only on the interior of the crystal. It is mainly used on synthetic sapphire crystals. Dubey & Schaldenbrand applies antireflection on both sides for all of the company's wristwatches such as this Aquadyn model.

AUTOMATIC WINDING

A rotating weight set into motion by moving the wrist winds the spring barrel via the gear train of a mechanical watch movement. Automatic winding was invented during the pocket watch era in 1770, but the breakthrough automatic winding movement via rotor began with the ball bearing Eterna-Matic in the late 1940s. Today we speak of unidirectional winding and bidirectionally winding rotors, depending on the type of gear train used. Shown is IWC's automatic Caliber 50611.

BALANCE

The beating heart of a mechanical watch movement is the balance. Fed by the energy of the mainspring, a tirelessly oscillating little wheel, just a few millimeters in diameter and possessing a spiral-shaped balance spring, sets the rhythm for the escape wheel and pallets with its vibration frequency. Today the balance is usually made of one piece of antimagnetic glucydur, an alloy that expands very little when exposed to heat.

BAR OR COCK

A metal plate fastened to the base plate at one point, leaving room for a gear wheel or pinion. The balance is usually attached to a bar called the balance cock. Glashütte tradition dictates that the balance cock be decoratively engraved by hand like this one by Glashütte Original.

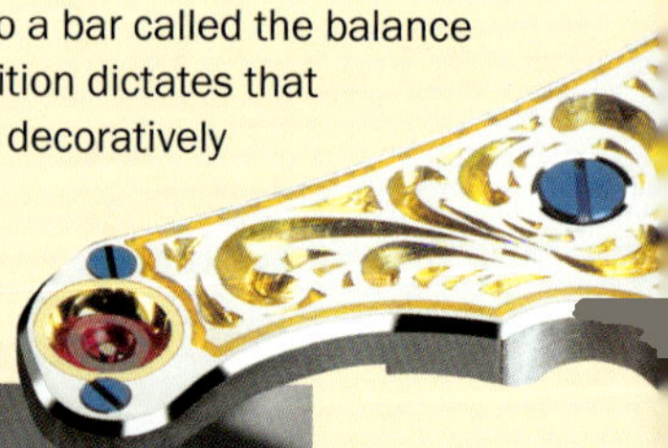

BEVELING

To uniformly file down the sharp edges of a plate, bridge, or bar and give it a high polish. The process is also called *anglage.* Edges are usually beveled at a 45° angle. As the picture shows, this is painstaking work that needs the skilled hands and eyes of an experienced watchmaker or *angleur.*

BRIDGE

A metal plate fastened to the base plate at two points leaving room for a gear wheel or pinion. This vintage Favre-Leuba movement illustrates the point with three individual bridges.

CARBON FIBER

A very light, tough composite material, carbon fiber is composed of filaments comprised of several thousand seven-micron carbon fibers held together by resin. The arrangement of the filaments determines the quality of a component, making each unique. Carbon fiber is currently being used for dials, cases, and even movement components.

CALIBER

A term, similar to type or model, that refers to different watch movements. Pictured here is Heuer's Caliber 11, the legendary automatic chronograph caliber from 1969. This movement was a coproduction jointly researched and developed for four years by Heuer-Leonidas, Breitling, and Hamilton-Büren. Each company gave the movement a different name after serial production began.

CHAMPLEVÉ

A dial decoration technique, whereby the metal is engraved, filled with enamel, and baked as in this cockatoo on a Cartier Tortue, enhanced with mother-of-pearl slivers.

CERAMIC

An inorganic, nonmetallic material formed by the action of heat and practically unscratchable. Pioneered by Rado, ceramic is a high-tech material generally made from aluminum and zirconia oxide. Today, it is used generally for cases and bezels and now comes in many colors.

CHRONOGRAPH

From the Greek *chronos* (time) and *graphein* (to write). Originally a chronograph literally wrote, inscribing the time elapsed on a piece of paper with the help of a pencil attached kind of hand. Today this term is used for watches that show not only the time of day, but also certain time intervals via independent hands that may be started or stopped at will. Stopwatches differ from chronographs because they do not show the time of day. This exploded illustration shows the complexity of a Breitling chronograph.

CHRONOMETER

Literally, "measurer of time." As the term is used today, a chronometer denotes an especially accurate watch (one with a deviation of no more than 5 seconds a day for mechanical movements). Chronometers are usually supplied with an official certificate from an independent testing office such as the COSC. The largest producer of chronometers in 2008 was Rolex with 769,850 officially certified movements. Chopard came in sixth with more than 22,000 certified L.U.C. mechanisms like the 4.96 in the Pro One model shown here.

COLUMN WHEEL

The component used to control chronograph functions within a true chronograph movement. The presence of a column wheel indicates that the chronograph is fully integrated into the movement. In the modern era, modules are generally used that are attached to a base caliber movement. This particular column wheel is made of blued steel.

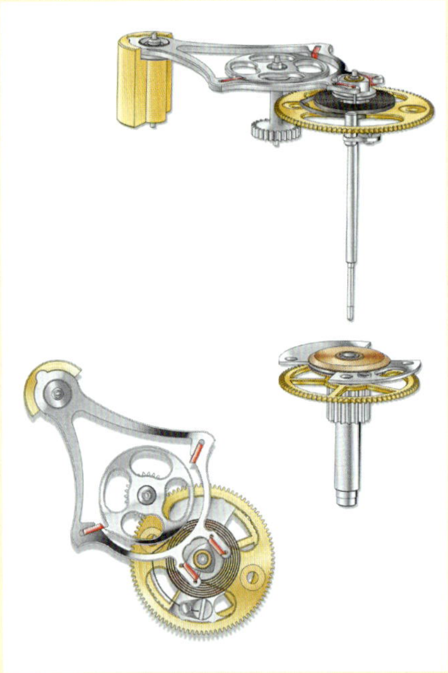

CONSTANT FORCE MECHANISM

Sometimes called a constant force escapement, it isn't really: in most cases this mechanism is "simply" an initial tension spring. It is also known in English by part of its French name, the *remontoir*, which actually means "winding mechanism." This mechanism regulates and portions the energy that is passed on through the escapement, making the rate as even and precise as possible. Shown here is the constant force escapement from A. Lange & Söhne's Lange 31—a mechanism that gets as close to its name as possible.

COSC

The Contrôle Officiel Suisse de Chronométrage, the official Swiss testing office for chronometers. The COSC is the world's largest issuer of so-called chronometer certificates, which are only otherwise given out individually by certain observatories (such as the one in Neuchâtel, Switzerland). For a fee, the COSC tests the rate of movements that have been adjusted by watchmakers. These are usually mechanical movements, but the office also tests some high-precision quartz movements. Those that meet the specifications for being a chronometer are awarded an official certificate as shown here.

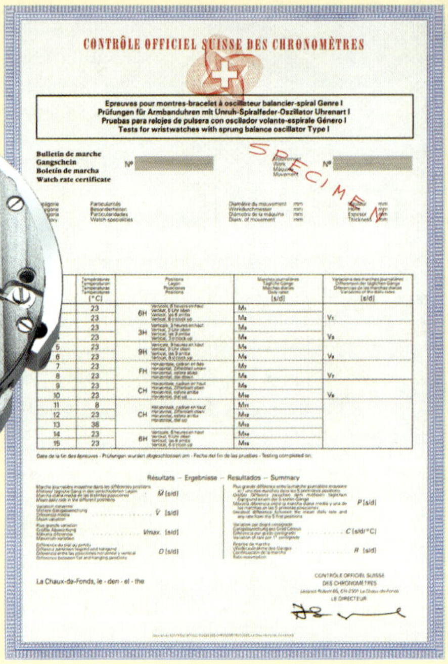

CÔTES DE GENÈVE

Also called *vagues de Genève* and Geneva stripes. This is a traditional Swiss surface decoration comprising an even pattern of parallel stripes, applied to flat movement components with a quickly rotating plastic or wooden peg. Glashütte watchmakers have devised their own version of *côtes de Genève* that is applied at a slightly different angle called Glashütte ribbing.

CROWN

The crown is used to wind and set a watch. A few simple turns of the crown will get an automatic movement started, while a manually wound watch is completely wound by the crown. The crown is also used for the setting of various functions, almost always including at least the hours, minutes, seconds, and date. A screwed-down crown like the one on the TAG Heuer Aquagraph pictured here can be tightened to prevent water entering the case or any mishaps while performing extreme sports such as diving.

EQUATION OF TIME

The mean time that we use to keep track of the passing of the day (24 hours evenly divided into minutes and seconds) is not equal to true solar time. The equation of time is a complication devised to show the difference between the mean time shown on one's wristwatch and the time the sun dictates. The Équation Marchante by Blancpain very legibly indicates this difference via the golden sun-tipped hand that also rotates around the dial in a manner known to watch connoisseurs as *marchant*. Other wristwatch models such as the Boreas by Martin Braun display the difference on an extra scale on the dial.

ESCAPEMENT

The combination of the balance, balance spring, pallets, and escape wheel, a subgroup which divides the impulses coming from the spring barrel into small, accurately portioned doses. It guarantees that the gear train runs smoothly and efficiently. The pictured escapement is one newly invented by Parmigiani containing pallet stones of varying color, though they are generally red synthetic rubies. Here one of them is a colorless sapphire or corundum, the same geological material that ruby is made of.

FLINQUÉ

A dial decoration in which a guilloché design is given a coat of enamel, softening the pattern and creating special effects, as shown here on a unique Bovet.

FLYBACK CHRONOGRAPH

A chronograph with a special dial train switch that makes the immediate reuse of the chronograph movement possible after resetting the hands. It was developed for special timekeeping duties such as those found in aviation, which require the measurement of time intervals in quick succession. A flyback may also be called a *retour en vol*. An elegant example of this type of chronograph is Corum's Classical Flyback Large Date shown here.

GEAR TRAIN

A mechanical watch's gear train transmits energy from the mainspring to the escapement. The gear train comprises the minute wheel, the third wheel, the fourth wheel, and the escape wheel.

GUILLOCHÉ

A surface decoration usually applied to the dial and the rotor using a grooving tool with a sharp tip, such as a rose engine, to cut an even pattern onto a level surface. The exact adjustment of the tool for each new path is controlled by a device similar to a pantograph, and the movement of the tool can be controlled either manually or mechanically. Real *guillochis* (the correct term used by a master of guilloché) are very intricate and expensive to produce, which is why most dials decorated in this fashion are produced by stamping machines. Breguet is one of the very few companies to use real guilloché on every one of its dials.

GLUCYDUR

Glucydur is a functional alloy of copper, beryllium, and iron that has been used to make balances in watches since the 1930s. Its hardness and stability allow watchmakers to use balances that were poised at the factory and no longer required adjustment screws.

INDEX

A regulating mechanism found on the balance cock and used by the watchmaker to adjust the movement's rate. The index changes the effective length of the balance spring, thus making it move more quickly or slowly. This is the standard index found on an ETA Valjoux 7750.

JEWEL

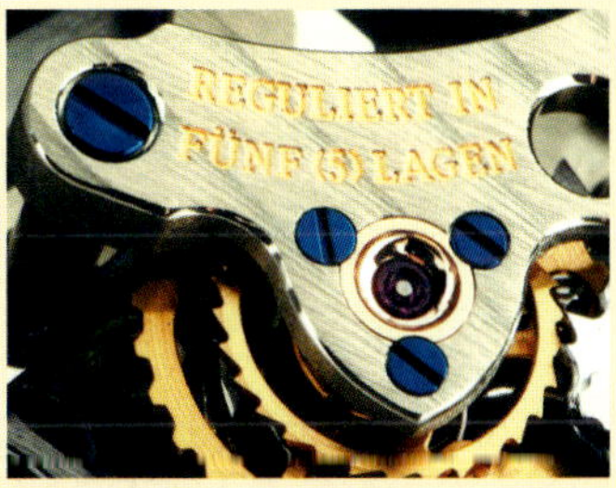

To minimize friction, the hardened steel tips of a movement's rotating gear wheels (called pinions) are lodged in synthetic rubies (fashioned as polished stones with a hole) and lubricated with a very thin layer of special oil. These synthetic rubies are produced in exactly the same way as sapphire crystal using the same material. During the pocket watch era, real rubies with hand-drilled holes were still used, but because of the high costs involved, they were only used in movements with especially quickly rotating gears. The jewel shown here on a bridge from A. Lange & Söhne's Double Split is additionally embedded in a gold chaton secured with three blued screws.

LIGA

The word LIGA is actually a German acronym that stands for lithography (*Lithografie*), electroplating (*Galvanisierung*), and plastic molding (*Abformung*). It is a lithographic process exposed by UV or X-ray light that literally "grows" perfect micro components made of nickel, nickel-phosphorus, or 23.5-karat gold atom by atom in a plating bath. The components need no finishing or trimming after manufacture.

LUMINOUS SUBSTANCE

Tritium paint is a slightly radioactive substance that replaced radium as luminous coating for hands, numerals, and hour markers on watch dials. Watches bearing tritium must be marked as such, with the letter T on the dial near 6 o'clock. It has now for the most part been replaced by nonradioactive materials such as Superluminova. Traser technology (as seen on these Ball timepieces) uses tritium gas enclosed in tiny silicate glass tubes coated on the inside with a phosphorescing substance. The luminescence is constant and will hold around twenty-five years.

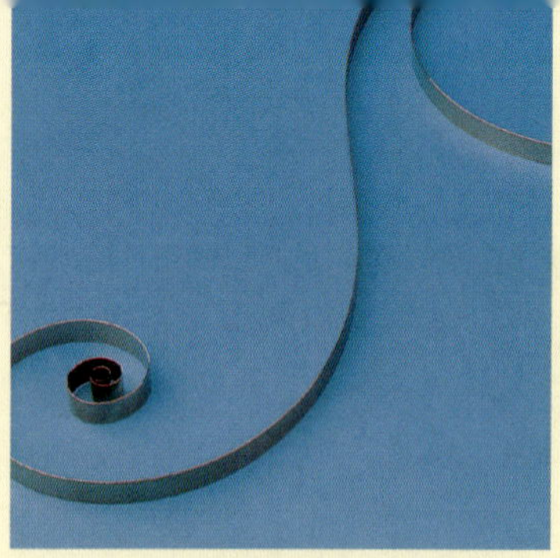

MAINSPRING

The mainspring, located in the spring barrel, stores energy when tensioned and passes it on to the escapement via the gear train as the tension relaxes. Today, mainsprings are generally made of Nivaflex, an alloy invented by Swiss engineer Max Straumann at the beginning of the 1950s. This alloy basically comprises iron, nickel, chrome, cobalt, and beryllium.

MINUTE REPEATER

A striking mechanism with hammers and gongs for acoustically signaling the hours, quarter hours, and minutes elapsed since noon or midnight. The wearer pushes a slide, which winds the spring. Normally a repeater uses two different gongs to signal hours (low tone), quarter hours (high and low tones in succession), and minutes (high tone). Some watches have three gongs, called a carillon. The Chronoswiss Répétition à Quarts is a prominent repeating introduction of recent years.

PERPETUAL CALENDAR

The calendar module for this type of timepiece automatically makes allowances for the different lengths of each month as well as leap years until the next secular year, which will occur in 2100. A perpetual calendar usually shows the date, month, and four-year cycle, and may show the day of the week and moon phase as well, as does this one introduced by George J von Burg at Baselworld 2005. Perpetual calendars need much skill to complete.

PERLAGE

Surface decoration comprising an even pattern of partially overlapping dots, applied with a quickly rotating plastic or wooden peg. (Here on the plates of Frédérique Constant's *manufacture* Caliber FC 910-1.

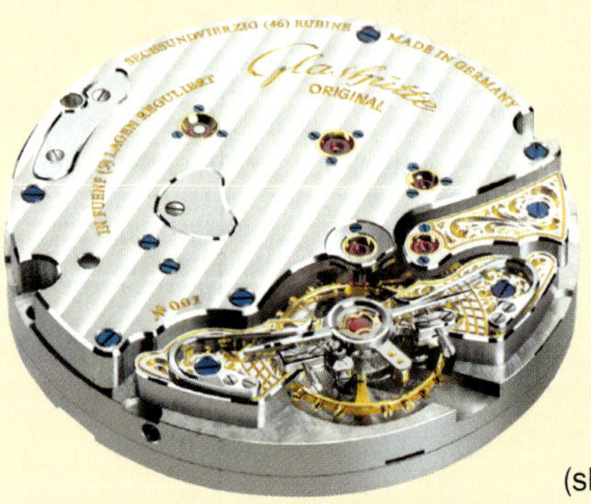

PLATE

A metal platform having several tiers for the gear train. The base plate of a movement usually incorporates the dial and carries the bearings for the primary pinions of the "first floor" of a gear train. The gear wheels are made complete by tightly fitting screwed-in bridges and bars on the back side of the plate. A specialty of the so-called Glashütte school, as opposed to the Swiss school, is the reverse completion of a movement not via different bridges and bars, but rather with a three-quarter plate. Glashütte Original's Caliber 65 (shown) displays a beautifully decorated three-quarter plate.

POWER RESERVE DISPLAY

A mechanical watch contains only a certain amount of power reserve. A fully wound modern automatic watch usually possesses between 36 and 42 hours of energy before it needs to be wound again. The power reserve display keeps the wearer informed about how much energy his or her watch still has in reserve, a function that is especially practical on manually wound watches with several days of possible reserve. The Nomos Tangente Power Reserve pictured here represents an especially creative way to illustrate the state of the mainspring's tension. On some German watches the power reserve is also displayed with the words "auf" and "ab."

PULSOMETER

A scale on the dial, flange, or bezel that, in conjunction with the second hand, may be used to measure a pulse rate. A pulsometer is always marked with a reference number—if it is marked with *gradué pour 15 pulsations*, for example, then the wearer counts fifteen pulse beats. At the last beat, the second hand will show what the pulse rate is in beats per minute on the pulsometer scale. The scale on Sinn's World Time Chronograph (shown) is marked simply with the German world *Puls* ("pulse"), but the function remains the same.

QUALITÉ FLEURIER

This certification of quality was established by Chopard, Parmigiani Fleurier, Vaucher, and Bovet Fleurier in 2004. Watches bearing the seal must fulfill five criteria, including COSC certification, passing several tests for robustness and precision, top-notch finishing, and being 100 percent Swiss made (except for the raw materials). The seal appears here on the dial of the Parmigiani Fleurier Tonda 39.

RETROGRADE DISPLAY

A retrograde display shows the time linearly instead of circularly. The hand continues along an arc until it reaches the end of its scale, at which precise moment it jumps back to the beginning instantaneously. This Nienaber model not only shows the minutes in retrograde form, it is also a regulator display.

ROTOR

The rotor is the component that keeps an automatic watch wound. The kinetic motion of this part, which contains a heavy metal weight around its outer edge, winds the mainspring. It can either wind unilaterally or bilaterally (to one or both sides) depending on the caliber. The rotor from this Temption timepiece belongs to an ETA Valjoux 7750.

SAPPHIRE CRYSTAL

Synthetic sapphire crystal is known to gemologists as aluminum oxide (Al_2O_3) or corundum. It can be colorless (corundum), red (ruby), blue (sapphire), or green (emerald). It is virtually scratchproof; only a diamond is harder. The innovative Royal Blue Tourbillon by Ulysse Nardin pictured here not only features sapphire crystals on the front and back of the watch, but also actual plates made of both colorless and blue corundum within the movement.

SCREW BALANCE

Before the invention of the perfectly weighted balance using a smooth ring, balances were fitted with weighted screws to get the exact impetus desired. Today a screw balance is a subtle sign of quality in a movement due to its costly construction and assembly utilizing minuscule weighted screws.

SEAL OF GENEVA

Since 1886 the official seal of this canton has been awarded to Genevan watch *manufactures* who must follow a defined set of high-quality criteria that include the following: polished jewel bed drillings, jewels with olive drillings, polished winding wheels, quality balances and balance springs, steel levers and springs with beveling of 45 degrees and *côtes de Genève* decoration, and polished stems and pinions. The list was updated in 2012 to include the entire watch and newer components. Testing is done on the finished piece. The Seal consists of two, one on the movement, one on the case. The pictured seal was awarded to Vacheron Constantin, a traditional Genevan *manufacture*.

SILICIUM/SILICON

Silicon is an element relatively new to mechanical watches. It is currently being used in the manufacture of precision escapements. Ulysse Nardin's Freak has lubrication-free silicon wheels, and Breguet has successfully used flat silicon balance springs.

SKELETONIZATION

The technique of cutting a movement's components down to their weight-bearing basic substance. This is generally done by hand in painstaking hours of microscopic work with a mini handheld saw, though machines can skeletonize parts to a certain degree, such as the version of the Valjoux 7750 that was created for Chronoswiss's Opus and Pathos models. This tourbillon created by Christophe Schaffo is additionally—and masterfully—hand-engraved.

SONNERIE

A variety of minute repeater that—like a tower clock—sounds the time not at the will of the wearer, but rather automatically (*en passant*) every hour (*petite sonnerie*) or quarter hour (*grande sonnerie*). Gérald Genta designed the most complicated sonnerie back in the early nineties. Shown is a recent model from the front and back.

SPLIT-SECONDS CHRONOGRAPH

Also known in the watch industry by its French name, the *rattrapante* (exploded view at left). A watch with two second hands, one of which can be blocked with a special dial train lever to indicate an intermediate time while the other continues to run. When released, the split-seconds hand jumps ahead to the position of the other second hand. The PTC by Porsche Design illustrates this nicely.

SPRING BARREL

The spring barrel contains the mainspring. It turns freely on an arbor, pulled along by the toothed wheel generally doubling as its lid. This wheel interacts with the first pinion of the movement's gear train. Some movements contain two or more spring barrels for added power reserve.

SWAN-NECK FINE ADJUSTMENT

A regulating instrument used by the watchmaker to adjust the movement's rate in place of an index. The swan neck is especially prevalent in fine Swiss and Glashütte watchmaking (here Lang & Heyne's Moritz model). Mühle Glashütte has varied the theme with its woodpecker's neck.

TACHYMETER

A scale on the dial, flange, or bezel of a chronograph that, in conjunction with the second hand, gives the speed of a moving object. A tachymeter takes a value determined in less than a minute and converts it into miles or kilometers per hour. For example, a wearer could measure the time it takes a car to pass between two mile markers on the highway. When the car passes the marker, the second hand will be pointing to the car's speed in miles per hour on the tachymetric scale.

TOURBILLON

A technical device invented by Abraham-Louis Breguet in 1801 to compensate for the influence of gravity on the balance of a pocket watch. The entire escapement is mounted on an epicyclic train in a "cage" and rotated completely on its axis over regular periods of time. This superb horological highlight is seen as a sign of technological know-how in the modern era. Harry Winston's Histoire de Tourbillon 4 is a spectacular example.

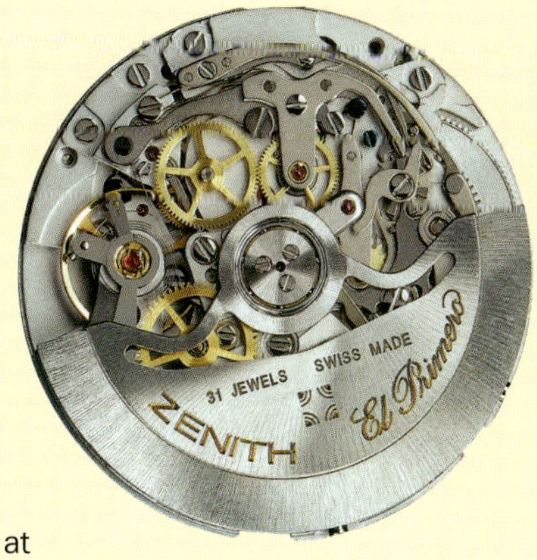

VIBRATION FREQUENCY (VPH)

The spring causes the balance to oscillate at a certain frequency measured in hertz (Hz) or vibrations per hour (vph). Most of today's wristwatches tick at 28,800 vph (4 Hz) or 21,600 vph (3 Hz). Less usual is 18,000 vph (2.5 Hz). Zenith's El Primero was the first serial movement to beat at 36,000 vph (5 Hz), and the Breguet Type XXII runs at 72,000 vph.

WATER RESISTANCE

Water resistance is an important feature of any timepiece and is usually measured in increments of one atmosphere (atm or bar, equal to 10 meters of water pressure) or meters and is often noted on the dial or case back. Watches resistant to 100 meters are best for swimming and snorkeling. Timepieces resistant to 200 meters are good for scuba diving. To deep-sea dive there are various professional timepieces available for use in depths of 200 meters or more. The Hydromax by Bell & Ross (shown) is water-resistant to a record 11,000 meters.

Editor-in-Chief: Peter Braun
Senior Editor: Marton Radkai
Copy Editor: Ashley Benning
Production and Layout: Muser Medien GmbH, Mannheim
Composition: Paul Aljian
Printed by: D + L Reichenberg GmbH

For more information about advertising, please contact:
Gary Girdvainis
25 Gay Bower Road
Monroe, CT 06468, USA
(203) 485-6276
gary@isochronmedia.com

For more information about book sales, please contact:
Customer Service Manager
Abbeville Press
137 Varick Street
New York, NY 10013, USA
1-800-ARTBOOK
www.abbeville.com

Printed and bound in Germany

ISBN 978-0-7892-1143-9

First Edition
10 9 8 7 6 5 4 3 2 1

Library of Congress Cataloging-in-Publication Data available upon request.

ISBN 978-0-7892-1143-9 U.S. $35.00
EAN
9 780789 211439 53500